ICE Themes Smart Concrete

The *ICE Themes* series showcases cutting-edge research and practical guidance in all branches
of civil engineering. Each title focuses on a key issue or challenge in civil engineering
and includes research from the industry's finest thinkers and influencers published through
the ICE Publishing programme. Themes in the series include climate change resilience,
advances in construction management, developments in renewable energy, and innovations
in construction materials plus many more.

Also available in the *ICE Themes* series:

ICE Themes Flood Resilience (2018)
Edited by Manuela Escarameia and Andrew Tagg. ISBN 978-0-7277-6393-8
ICE Themes Low Carbon Concrete (2019)
Edited by Ravindra K Dhir OBE and Kevin Paine. ISBN 978-0-7277-6459-1
ICE Themes Recycled Aggregates: Use in Concrete (2019)
Edited by Ravindra K Dhir OBE and Chao Qun Lye. ISBN 978-0-7277-6463-8
ICE Themes Renewable Energy: Geothermal Energy, Heat Exchange Systems and Energy Piles (2018)
Edited by William Craig and Kevin Gavin. ISBN 978-0-7277-6398-3
ICE Themes Renewable Energy: Wind Turbine Foundations (2018)
Edited by Kenneth Gavin and William Craig. ISBN 978-0-7277-6396-9
ICE Themes Self-Compacting Concrete (2019)
Edited by Ravindra K Dhir OBE and Chao Qun Lye. ISBN 978-0-7277-6461-4

ICE Themes
Smart Concrete

Edited by Ravindra K Dhir OBE and Kevin Paine

Published by ICE Publishing, One Great George Street, Westminster, London SW1P 3AA.

Full details of ICE Publishing representatives and distributors can be found at: www.icebookshop.com/bookshop_contact.asp

Other titles by ICE Publishing:

Steel-Concrete Composite Bridges, Second edition.
David Collings. ISBN 978-0-7277-5810-1
Deterioration and Maintenance of Pavements
Derek Pearson. ISBN 978-0-7277-4114-1
Finite-Element Design of Concrete Structures, Second edition
Guenter A. Rombach. ISBN 978-0-7277-4189-9

www.icebookshop.com

A catalogue record for this book is available from the British Library

ISBN 978-0-7277-6457-7

© Thomas Telford Limited 2019

ICE Publishing is a division of Thomas Telford Ltd, a wholly-owned subsidiary of the Institution of Civil Engineers (ICE).

Cover photo: L'Umbracle building, with El Pont de l'Assut de l'Or in the background, City of Arts and Sciences, Valencia, Spain. Digital-Fotofusion Gallery/Alamy Stock Photo

Commissioning Editor: James Hobbs
Production Editor: Madhubanti Bhattacharyya
Marketing Specialist: April Asta Brodie

Typeset by The Manila Typesetting Company
Index created by The Manila Typesetting Company
Printed and bound in Great Britain by TJ International Padstow

Contents

Foreword

This themed book covers important developments in the field of smart concrete as published in ICE journals in the past decade. The concept of smart concrete is new and it continues to evolve as further developments emerge. As yet there is no accepted definition for smart concrete, and it is sometimes used synonymously with the terms intelligent concrete and autonomic concrete. What is agreed is that smart concrete needs to be programmed, through changes in its composition, microstructure or the addition of actuating components, to respond to and control its own actions. Smart concrete cuts across a number of fields, and a recent review of the subject area suggests that it covers topics as diverse as self-sensing, self-healing, self-adjusting, self-heating, self-cleaning, electromagnetic shielding/absorbing, energy harvesting, light- transmitting, and aircraft arresting properties [1].

In putting this theme book together, the editors have considered that smartness of concrete can manifest itself in several aspects, such as the ability of concrete to detect (sense) changes within itself and be able to respond to them effectively. Smartness can also relate to the ability of the concrete to be able to carry out procedures and fulfil requirements that would normally need some form of human intervention – for example curing, healing, and damage detection.

Whilst smart concrete is deemed to have properties that normal concrete does not have, the separation between what may be considered normal and what is smart, is debateable and potentially controversial. Nonetheless, it is necessary to distinguish between what we have traditionally expected from concrete (workability, strength, durability) with what additional properties may be needed for concrete to achieve further aims of sustainability and greater resilience. Currently, it is generally recognised that concrete is perhaps not as durable as it should be. Indeed, an on-going issue with concrete has been the need for human intervention to look after it in operation. This creates costs that are often un-budgeted. However, because concrete is subject to a range of environments and exposure conditions, and because these can change from what was considered at the design stage, there are perhaps limits to what we can do with the concrete mix design to improve this aspect. This will become more prevalent as climate change takes hold. Should we design concrete for environments as they are now in the 2020s or for what they are likely to be in 2060s, when any

newly constructed concrete now will still be in service?
Or should we design concrete that has the ability to adapt
itself? So-called smart concrete!

The anticipated need for smart concrete drives this entire
research area. The number of papers submitted to ICE
Publishing each year is symbolic of the vast importance of
concrete as a construction material, but also of the need to
better understand it, amend it or change it. It is likely that
what we have highlighted as smart concrete in this theme
book, will in a few years' time be considered mainstream
and be an expected requirement of normal concrete. We shall
see! Indeed, this is a fast-evolving field. In putting this book
together, a search of all ICE publications containing cement
and concrete related papers was undertaken back to the
1950s. However, in terms of what may now be considered
'smart', this book includes only one paper published prior
to 2010.

The book starts with two papers on curing. In most treatises
on smart concrete, this topic is not covered. However, poor
curing remains a critical aspect of many of our problems
with concrete infrastructure. Concrete will of course cure
well in the right environment (generally under water) and
this could be regarded as a natural ability of concrete.
However, mostly, concrete is not exposed to such conditions
and consequently it is unable to cure well. Humans therefore
have to manually assist in this action. Concrete that does
not require such primary assistance can be regarded in this
aspect as smart. The first paper describes the work to create
concrete with improved water-retention properties.
Consequently, water does not evaporate, and remains
available to hydrate the cement. In more recent work, as
described in the second paper, an alternative system is
utilised, but one that can be regarded as providing greater
'smartness'. Here the concrete is provided with an additional
internal reservoir of water that it can choose to use, or not,
depending on requirements, for example, where water
evaporation is great, this reservoir will be used to replace
that water and maintain effective curing.

The next section covers a form of concrete that is at the
cutting edge of current research: self-healing concrete.
Repair and maintenance accounts for considerable amount
of expenditure. Much of this is due to the occurrence of
cracking, especially microcracking, which provides pathways
for harmful moisture, oxygen and chlorides to the

reinforcing bars. This initiates corrosion and leads to
further cracking and spalling of concrete in the cover zone.
Traditionally, this cracking has been accepted as a fait
accompli of utilising the material, and repeated cycles of
human mediated repair accepted as the norm. However, in
practice due to problems in detecting cracking, and meeting
the cost of repair, the lifetime of the material (and structure)
are compromised. The section starts with a paper from 2011
that provides a good summary of important terms in the
field and initial solutions to providing an autonomic smart
self-healing concrete. A later paper by **Al-Tabbaa _et al._**
provides a recent update on the technologies available
with specific reference to the leading UK research in
self-healing concrete.

In addition to these autonomic approaches, the ability of
concrete cracks to close autogenously is an important feature
and one that has been studied for some time (e.g. see [2]).
In modern terms this autogenous healing may be enhanced
when fibres are added. This is particularly the case for
engineered cementitious composites which are specifically
designed to develop multiple small cracks (of less than
0.1 mm) rather than large cracks. Self-healing of these
cracks within days and at worst weeks is also discussed
in the papers in this section. Notwithstanding the progress
made within the field of self-healing of concrete, more
research is required; for example, to determine whether
concrete that has self-healed does actually regain full
functionality for resistance to chloride ingress and
freeze/thaw. This is discussed in the final paper of
this section.

Following this, we look at the addition of phase-change
materials and carbon to concrete to improve its properties.
Phase change materials allow concrete to act as latent heat
storage units and allow control of the interior temperature
rise in mass concrete. Whilst the degree of smartness in the
paper given here is limited, it is a starting point to the use
of phase change materials to smartly control heat flow in
and out of concrete walls to regulate internal building
temperatures. This will play an important role in modern
sustainable building design to reduce heating at night and
the need for air-conditioning during the day.

To achieve fully self-sensing concrete (the topic of the next
section), it may be necessary to modify the cementitious
matrix to increase its electrical conductivity. This can be

done by utilising highly conductive materials or the piezoelectric properties of carbon nanotubes. However, a significant problem, and one that also applies to self-healing concrete above, is how to embed the components that provide the 'smartness' into concrete without comprising the inherent properties (workability, strength, and hardness). This issue is looked at in detail with specific reference to dispersion of fine carbon materials in concrete.

The final section relates to monitoring and sensing of concrete. A lot of initial research in this area was related to the use of sensors in concrete to provide messages and signals to engineers to permit them to monitor the health of a structure and to allow intervention where necessary. However, increasingly the sensors are being used to provide an autonomic response from the concrete itself, either via computer-based algorithms or by triggering a change in the material itself. This latter aspect should take us further towards concrete that is fully 'aware' of its own condition and even able to select and choose how it decides to respond to changes in its environment and condition. In this way we would most probably move from 'smartness' towards 'intelligence'.

This book has been compiled from papers previously published by ICE Publishing. Unsurprisingly, most of the papers have been taken from *Magazine of Concrete Research* and *Advances in Cement Research*, demonstrating the importance of these journals in transmitting concrete and cement research.

We are grateful to the contributing authors for their insights and hope that the readers will find this book a useful reference and source of motivation in their quest for knowledge in the field of smart concrete, as it continues to evolve into new areas.

Ravindra K Dhir OBE and Kevin Paine

REFERENCES

[1] B. Han *et al.*, "Smart concretes and structures: A review," *J. Intell. Mater. Syst. Struct.*, vol. 26, no. 11, pp. 1303–1345, 2015.
[2] R. K. Dhir, C. M. Sangha, and J. G. L. Munday, "Strength and deformation properties of autogenously healed mortars," *ACI J.*, vol. 70, no. 3, pp. 231–236, 1973.

About the editors

Professor Ravindra K Dhir OBE, an Emeritus Professor of Concrete Technology at the University of Dundee in the UK, also holds the positions of adjunct professor at Trinity College Dublin Ireland and honorary professor at the University of Birmingham UK. He remains actively engaged with his teaching, research and publishing work, as well his consultancy practice, *Applying Concrete Knowledge*. As a founding Director of Concrete Technology Unit, at Dundee (1988–2008), his hallmark was the development of a world renowned Centre of Excellence with extensive research facilities, as well as working very closely with industry in research and its dissemination into practice. He won numerous recognitions, awards and honours including the Order of the British Empire (OBE) from the Queen, and Honorary Fellowships of the Institute of Concrete Technology, UK and Indian Concrete Institute. He was Chairman of Concrete Society, Scotland (1986–87) and President of Concrete Society, UK (2009–2010). He serves on the Editorial Board of the *Magazine of Concrete Research* and is the Chairman of UK-India Education Research Initiative Concrete Congress.

Dr Kevin Paine graduated with a PhD in civil engineering from the University of Nottingham in 1998 for work on prestressed fibre reinforced concrete. He joined the Concrete Technology Unit at the University of Dundee as a Research/ Teaching Fellow (later promoted to Lecturer) where he published widely on the use of industrial by-products and recycled materials as cements and aggregates. In 2007 he was appointed Senior Lecturer (translated to Reader in 2015) within the BRE Centre for Innovative Construction Materials at the University of Bath. His most recent research has concentrated on development of nanotechnologically enhanced cements and self-healing and self-sensing concretes. He sits on a number of RILEM technical committees and is an editorial board member for the Institution of Civil Engineers' Construction Materials journal. He has published, lectured and examined on cement science and concrete technology around the world.

Self-curing

Dhir and Paine
ISBN 978-0-7277-6457-7
https://doi.org/10.1680/icetsc.64577.003
ICE Publishing: All rights reserved

Chapter 1

Paraffin wax as an internal curing agent in ordinary concrete

Madduru Sri Rama Chand
Research Scholar, Department of Civil Engineering, National Institute of Technology, Warangal, India

Pollapothu Swamy Naga Ratna Giri
Research Scholar, Department of Civil Engineering, National Institute of Technology, Warangal, India

Garje Rajesh Kumar
Professor, Department of Civil Engineering, National Institute of Technology, Warangal, India

Pancharathi Rathish Kumar
Associate Professor, Department of Civil Engineering, National Institute of Technology, Warangal, India

Curing of concrete involves maintaining satisfactory moisture content during early stages to develop the desired properties. Properly cured concrete has improved durability and surface hardness, and is less permeable. Prevention of loss of moisture is important not only for strength development but also to prevent plastic shrinkage, for decreased permeability and to improve resistance to abrasion. Good and complete curing is not always practical for several reasons, particularly in higher grade concretes. Using self-curing agents can solve this problem. The concept of self-curing agents is to reduce water evaporation from concrete, and hence increase its water retention capacity compared to conventional curing. Several materials, including polymeric glycol and paraffin wax, can act as self-curing compounds. This study investigates the role of paraffin wax as a self-curing agent and compares this with the effect of different curing regimes simulating traditional methods of curing. The parameters include grade of concrete, type and dosage of paraffin wax, curing conditions and age of curing. Weight loss and compressive strength are determined as a performance benchmark for the investigated curing compounds. It is found that the lower dosage (0·1%) liquid paraffin wax compounds act as the best curing compounds in higher grade concretes.

Introduction

Water is the maximum utilised commodity and because of this the day-by-day level of the water table is going down. If water has to be purchased for construction works, the cost of construction rises much higher. Also, in case of concreting works done at heights, vertical members, sloped roofs and pavements, continuous curing is very difficult. Where the thickness of concreting is large, the percolation of water in the concrete is difficult, especially in the case of high-strength concrete. In the case of high-performance concrete/self-compacting concrete, where the surface pores and mixing water are minimised, complete curing of cement particles does not take place (Cano Barrita *et al.*, 2003; Debashis and Abhijit, 2012).

Internal curing is especially beneficial in low water–cement ratio (w/c) based concretes because of the chemical shrinkage that accompanies Portland cement hydration and the low permeability of calcium silicate hydrates. Because water that is chemically bound and absorbed by cement hydration products has a specific volume less than that of bulk water, a hydrating cement paste will imbibe water from the available source (Jianhua and Yingshu, 2013). In higher w/c concretes, this water can be, and often is, supplied by external curing, but in concretes with low w/c ratio, the permeability of the concrete quickly becomes too low to allow the effective transfer of water from the external surface to the concrete interior. This is one of the justifications for internal curing. Additional water that can be distributed somewhat uniformly throughout the concrete will be more readily able to reach un-hydrated cement. This can be achieved by using pre-soaked lightweight aggregates (LWAs) or super-absorbent polymer (SAP). Use of LWA/SAP in concrete involves an additional 'carbon footprint', owing to their manufacturing process. Furthermore, the strength of concrete is likely to be reduced because of the low mechanical strength of LWA/SAP (Jiajun *et al.*, 2006; Zhutovsky *et al.*, 2002).

In the above-described circumstances, internal curing can be effectively accomplished by the use of water-sealing materials, namely, polyethylene glycol, paraffin wax, poly acrylic acid and so on. Water-soluble polymers can be used as self-curing agents and are likely to reduce the water evaporation from concrete to increase its water-retention capacity compared to conventional concrete (Lura, 2003). Concrete incorporating self-curing agents will represent a new trend in concrete construction in the current millennium. As is already known, curing of concrete plays a major role in developing an improved concrete microstructure and pore structure, which indirectly improves its durability and performance (Bentz and Snyder, 1999). Bentz *et al.* (2002) studied the addition of LWA sand to provide internal curing in cement mortars and examined the interfacial transition zone (ITZ). They concluded that, owing to the presence of LWA sand during the internal curing regime, there is improvement in the ITZ in terms of strength and durability (Bentz, 2009). Cussion and Hoogeveen (2008) conducted experiments on pre-soaked LWA sand to study autogenous shrinkage cracking. It was noted that there is an improvement in tensile cracking behaviour upon use of self-curing compounds. Jensen and Freiesleben (2001, 2002) addressed the curing of concrete using SAPs and explained the effectiveness of curing with different w/c ratios. The same is stressed in state-of-the-art reports on internal curing of concrete using LWAs and SAPs (Bentz and Weiss, 2011; Machtcherine and Reinhardt, 2012; Poursaee and Hansson, 2010).

Because of the non-availability of both LWAs and SAPs, and bearing in mind the technological difficulties in producing these materials, a method of curing concrete, namely, internal sealing – which does not need an externally applied curing, but is not based on adding water to the concrete – has been suggested as a viable alternative. Dhir *et al.* (1994, 1995) studied the strength and durability performance of self-curing concrete using hydrophilic chemicals. Adding hydrophilic chemicals to the water when mixing water with concrete will reduce water evaporation as the concrete is exposed to air-drying (Dhir *et al.*, 1998). Liang and Sun (2002) carried out work on internal curing of concrete using polyethylene glycol and paraffin wax. Wen-Chen (2008) experimented with the use of poly acrylic acid and polyvalent alcohol as self-curing agents and evaluated the optimum dosages in concrete. El-Dieb (2007), in his investigation, evaluated the use of water-soluble polymeric glycol as a self-curing agent and compared water retention and hydration of concrete containing self-curing agents with conventional concrete (El-Dieb *et al.*, 2011).

Experimental approach and materials

Of the several methods discussed in literature, those using compounds which act as sealing materials seem to be more effective. This enables no loss of water from the concrete. In the present study, experimental work was carried out to establish the suitability of a curing compound and its dosage in different grades of concrete. The investigation was done considering concrete mixes (1:1·4:2·6) with w/c ratios 0·35 and 0·45 designated as 'A' and 'B' grades. The dosage considered for different types of water-sealing compounds used in the study were 0%, 0·1%, 1·0% and 2·0% by weight of cement. The types of curing investigated in the study were indoor curing, water immersion and internal curing, that is curing by water-sealing compounds. A comparison was made with specimens cured under indoor conditions and conventional curing.

Materials used

The different materials used in this investigation are discussed below.

Cement

Cement used in the investigation was 53 grade ordinary Portland cement conforming to IS 12269 (BIS, 1987). The specific gravity of the cement was 3·14 and its specific surface area was 225 m^2/g; its initial and final setting times were 40 min and 560 min respectively.

Fine aggregate

The fine aggregate conformed to Zone-2 according to IS 383 (BIS, 1970). The fine aggregate used was obtained from a nearby river source. The specific gravity was 2·65 and the bulk density of the sand was 1·45 g/cm^3.

Coarse aggregate

Crushed granite was used as coarse aggregate. The coarse aggregate was obtained from a local crushing unit and had nominal size of 20 mm; it was well-graded aggregate according to IS 383 (BIS, 1970). The specific gravity was 2·8 and the bulk density was 1·5 g/cm^3.

Water

Potable water was used in the experimental work for both mixing and curing companion specimens.

Superplasticiser

High-range water-reducing admixtures conforming to ASTM C494 (ASTM, 2005), commonly called superplasticisers, were used to improve the flow or workability for decreased

Table 1 Physical and chemical properties of paraffin wax

Light liquid paraffin wax		Solid paraffin wax	
Flash point: °C	> 180	Flash point: °C	> 204
Appearance	Clear colourless liquid.	Appearance	White
Specific gravity: g/ml	0·845 at 15°C	Specific gravity: g/ml	0·96 at 15°C
Vapour density (air = 1)	> 1	Physical state	Solid – pastille form
Vapour pressure	< 0·01 mmHg at 2·0°C	Boiling point	343°C
		Melting point	58·9–62·2°C
Solubility in water	Insoluble Soluble in petroleum solvents	Solubility in water	Negligible

w/c ratio, without sacrificing compressive strength. These admixtures, when they disperse in cement agglomerates, significantly decrease the viscosity of the paste, forming a thin film around the cement particles.

Water-sealing chemicals

Liquid paraffin wax and solid paraffin wax were used in the study. The chemicals were mixed with water thoroughly before mixing the water into the concrete. Solid paraffin wax was melted at 60°C before adding to the water. The details of the physical properties of the paraffin wax compounds are shown in Table 1.

Nomenclature of specimens

The specimens were designated according to grade, type of paraffin wax and dosage. Typically 'A' (1:1·4:2·6/0·35) represents higher grade, while 'B' (1:1·4:2·6/0·45) represents lower grade concretes. Solid paraffin wax is designated as 'S', while 'V' denotes liquid paraffin wax. 'I' represents indoor curing with 0% dosage and 'W' represents conventional/water curing with 0% dosage. The percentages of dosage are taken as 0, 0·1, 1·0 and 2·0% by weight of cement. For example AV-0·1 represents a specimen with higher grade containing the curing compound of liquid paraffin wax at a dosage of 0·1%.

Experimental work

The experimental work involves a comparison of broadly three different curing conditions: indoor curing (I), water curing (W) and self-curing compounds liquid paraffin wax (V) and solid paraffin wax (S) at different dosages. The effectiveness of the chosen compounds can be compared based on water retention and compressive strength. These hydrophilic self-curing agents are supposed to reduce the evaporation of water and increase the water retention capacity for the hydration process. Comparison of these self-curing agents is carried out using indoor curing (I) and conventional wet curing (W).

Water retention test

The resource of the water molecules in concrete, particularly when subjected to indoor curing, is extremely important for the curing of concrete. Curing compounds added during the mixing time act as internal sealing agents, by filling cracks, voids, decreasing self-desiccation and progressing the hydration of concrete. The retention of water in concrete was monitored by weighing the concrete cubes at regular intervals. The weight of three similar specimens was taken on an electronic balance of accuracy 0·1 g and the average was taken for further analysis.

Compressive strength test

Standard cube specimens were cast and tested for compression using a Tenius Olsen testing machine. The maximum load that the specimen could sustain was recorded. The rate of loading was adopted as per IS 516 (BIS, 1956).

Results and discussion

Water retention

Water retention is an important parameter of investigation as maintaining satisfactory moisture content, particularly at early ages, is important for hydration. In this study a comparison was made of water retention chemicals, namely, liquid paraffin wax and solid paraffin wax.

The initial weight of each specimen type was taken as the reference value. The weight loss with respect to time was observed. The difference between the weight at any age and the initial weight (reference value) is the weight loss and this is plotted with respect to the age of the specimen. Figures 1–4 show plots of the average weight loss of liquid paraffin wax and solid paraffin wax in two grades of concrete, namely, A (60 MPa) and B (40 MPa)

Figure 1 Average weight loss in A grade with liquid paraffin wax

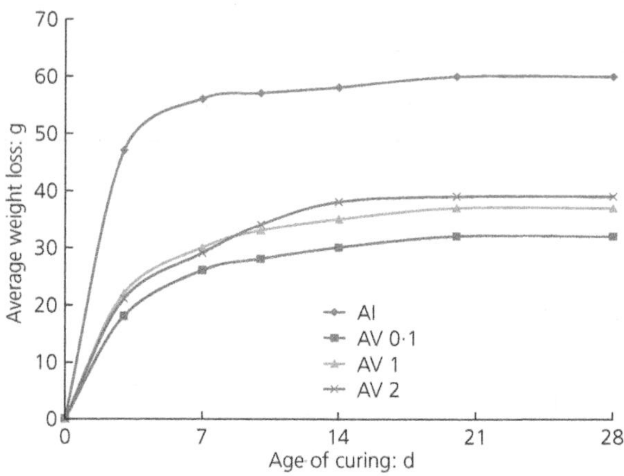

Figure 2 Average weight loss in B grade with liquid paraffin wax

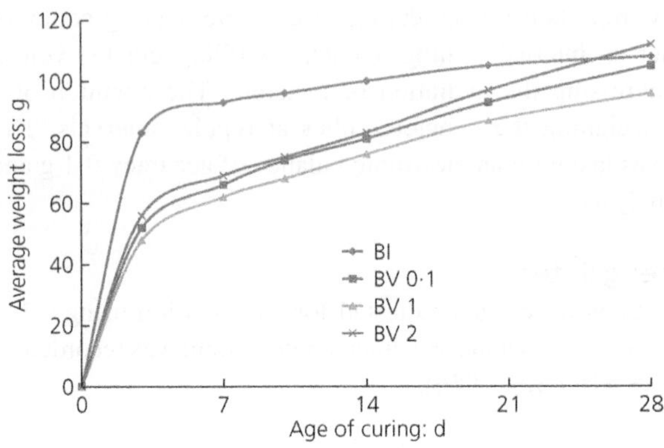

Figure 3 Average weight loss in A grade with solid paraffin wax

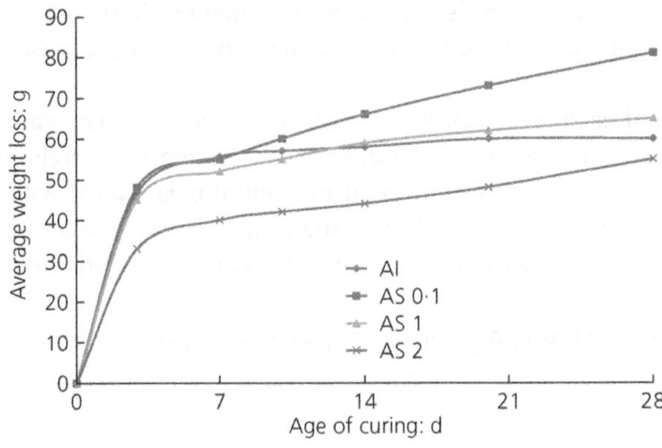

Figure 4 Average weight loss in B grade with solid paraffin wax

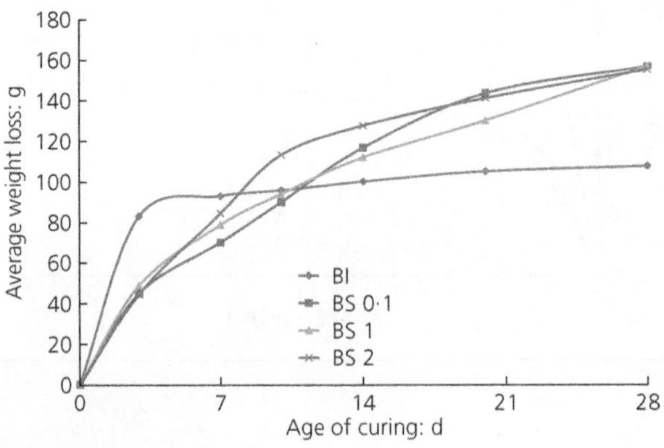

between 0 and 28 d. The final weight loss of all the specimens (i.e. after 28 d) is recorded as shown in Table 2. It can be observed that there is a rapid moisture loss up to 7 d and the percentage weight loss decreases as the curing progresses.

Figures 5–8 show the details of the relative weight loss ratio with respect to reference specimens, that is, indoor curing concrete (0% water sealing compound). It can be noted from Figures 5 and 6 that the ratio of weight loss is minimum in liquid paraffin wax specimens at 0·1% dosage in the case of higher grade concrete as compared to the reference. This be attributed to lower w/c ratio and dense pore structure. It can hence be concluded that for higher grade concretes using liquid paraffin wax at lower dosages is beneficial.

From Figure 6, it can be noted that, at higher w/c ratio, because of weak pore structure the replenishment of a larger dosage of liquid paraffin wax (i.e. 1·0%) is needed. Figures 7 and 8 show a comparison between the use of solid paraffin wax in higher and lower grades of concrete. It can be noted in this case that use of solid paraffin wax is not beneficial for either the higher grade (A) or the lower grade (B).

Table 2 28 d average weight loss

Dosage: %	28 d average weight loss: g					
	AI	AV	AS	BI	BV	BS
0	60	–	–	108	–	–
0·1	–	32	81	–	105	157
1·0	–	37	65	–	96	157
2·0	–	39	55	–	112	155

Figure 5 Relative weight loss ratio in A grade liquid paraffin wax

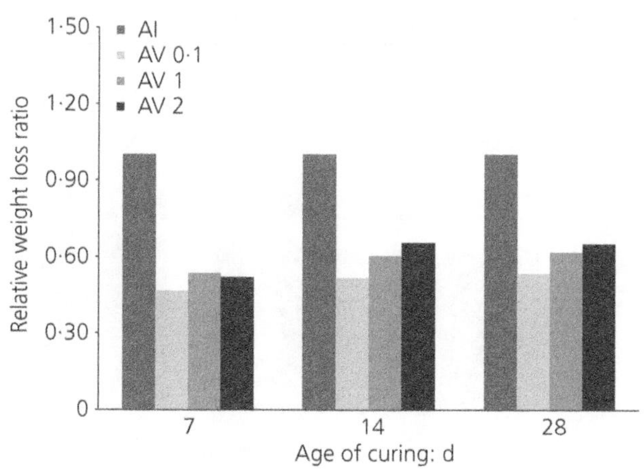

Figure 6 Relative weight loss ratio in B grade liquid paraffin wax

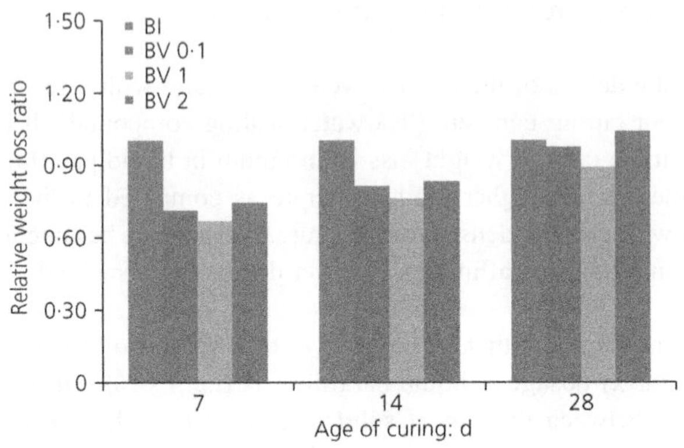

Figure 7 Relative weight loss ratio in A grade solid paraffin wax

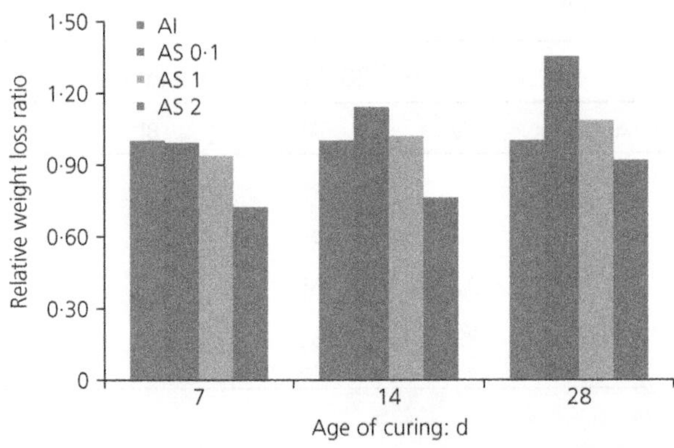

Figure 8 Relative weight loss ratio in B grade solid paraffin wax

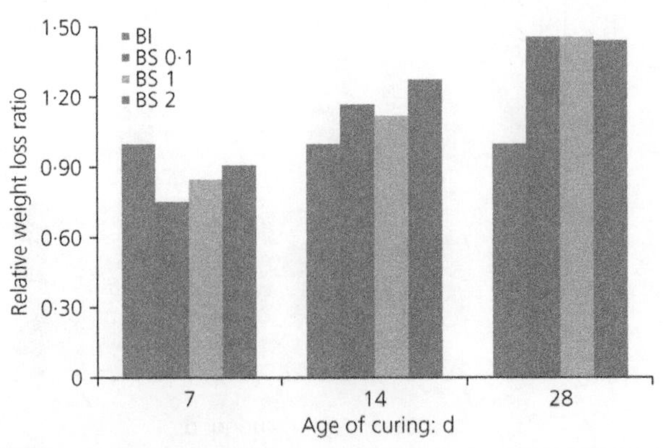

In all of the above cases, addition of liquid paraffin wax gave better performance than specimens subjected to indoor curing (i.e. no curing). It can be noted from Table 2 that the loss of water with identical w/c rato (in both grades A and B) is low with use of liquid paraffin wax in higher grade concretes as compared to lower grade concretes. Hence from the water retention test, it can be concluded that addition of liquid paraffin wax is more beneficial for higher grade concretes than lower grade concretes.

Variation of compressive strength

The compressive strength test was done at the end of 7 d, 14 d and 28 d, with an average of three specimens for each curing period. Figures 9 and 10 show the details of average compressive strength values of higher and lower grade concretes (60 MPa and 40 MPa) with liquid

Figure 9 Age plotted against compressive strength for A grade with liquid paraffin wax

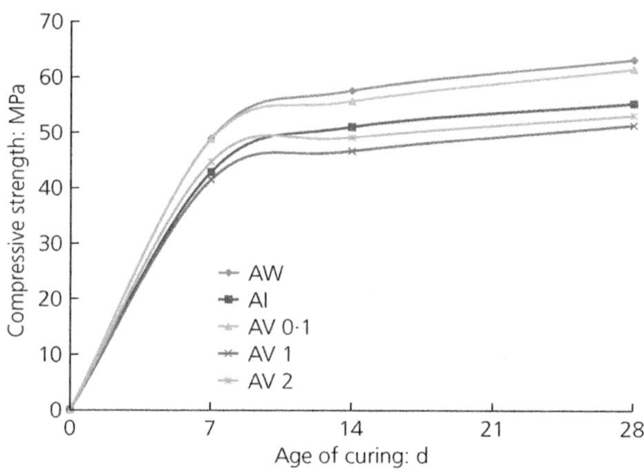

Figure 10 Age plotted against compressive strength for B grade with liquid paraffin wax

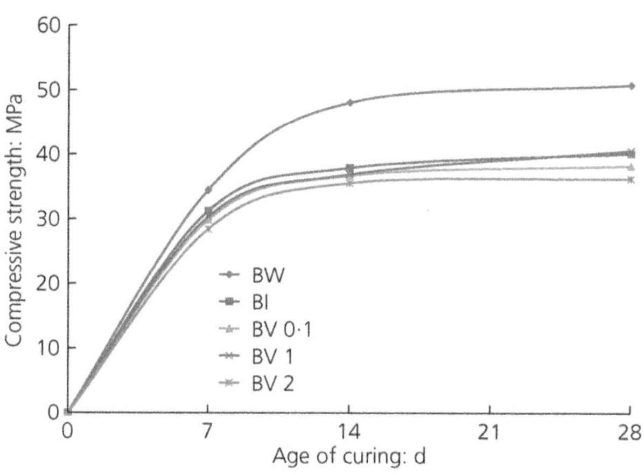

paraffin wax, while Figures 11 and 12 show the details of compressive strength values at different ages for grades A and B (60 MPa and 40 MPa) with solid paraffin wax. The details of compressive strength after 28 d are given in Table 3.

It can be noted in general that the use of liquid paraffin wax is beneficial in higher grades of concrete compared to lower grades. All the specimens could almost reach the strength attained through wet curing. It can also be observed that with 0·1% liquid paraffin wax at an age of 28 d the compressive strength is almost equal to the conventional wet curing specimen, with only marginal difference. Hence, it can be concluded that in applications where water curing is not possible, self-curing compounds replenish the loss of moisture indirectly, thereby aiding the hydration process.

Figure 11 Age plotted against compressive strength for A grade with solid paraffin wax

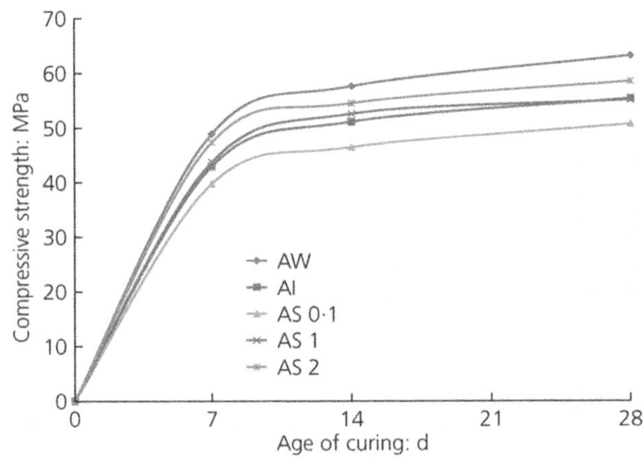

Figure 12 Age plotted against compressive strength for B grade with solid paraffin wax

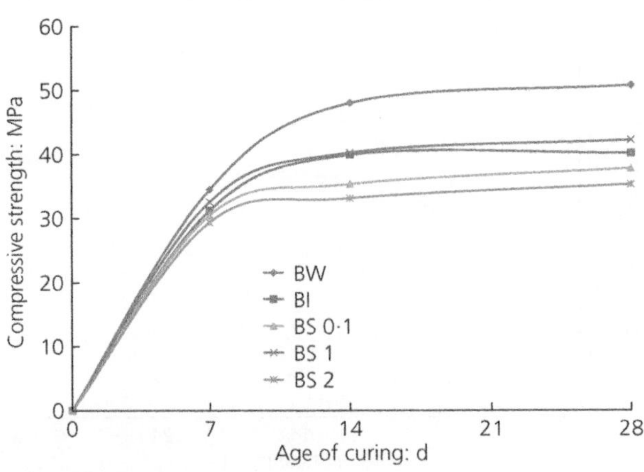

Table 3 28 d compressive strength (MPa) of concrete

%	AW	AI	AV	AS	BW	BI	BV	BS
0	63·2	55·36	–	–	50·75	40·14	–	–
0·1	–	–	61·50	50·75	–	–	38·23	37·78
1	–	–	51·43	55·13	–	–	40·56	42·16
2	–	–	53·19	58·54	–	–	36·24	35·27

It is also observed that self-curing compounds are not at all beneficial in lower grade concrete (B). In fact, for certain dosages of self-curing compounds, for both liquid paraffin wax (V) and solid paraffin wax (S), there is a decrease in compressive strength.

Conclusions

(*a*) The effectiveness of self-curing concrete is affected by w/c ratio and percentage dosages of the self-curing agent.

(*b*) The water retention of concrete specimens cured by self-curing compound liquid paraffin wax is higher compared to indoor cured specimens.

(*c*) There is a significant effect relating to the compressive strength of concrete with low w/c ratio owing to change in curing regime.

(*d*) In terms of water retention of concrete with lower w/c ratio incorporating liquid paraffin wax, lower dosage (0·1%) is more beneficial.

(*e*) In terms of water retention of concrete with higher w/c ratio, incorporating a self-curing agent is not beneficial.

(*f*) The compressive strength of concrete with low w/c ratio and lower dosage (0·1%) of liquid paraffin wax is high compared to other dosages (1·0%, 2·0%).

(*g*) Solid paraffin wax does not seem to provide a benefit as a self-curing agent. It could not satisfy water retention or compressive strength requirements.

REFERENCES

ASTM (2005) ASTM C494: Standard specification for chemical admixtures for concrete. ASTM, West Conshohocken, PA, USA.

Bentz DP, Geiker M and Jensen OM (2002) On the mitigation of early age cracking – self-desiccation and its importance in concrete technology. *Proceedings of the 3rd International Research Seminar, Lund, Sweden,* pp. 195–204.

Bentz DP and Snyder KA (1999) Protected paste volume in concrete extension to internal curing using saturated lightweight fine aggregate. *Cement and Concrete Research* **29(11)**: 1863–1867.

Bentz DP (2009) Influence of internal curing using lightweight aggregates on interfacial transition zone percolation and chloride ingress in mortars. *Cement and Concrete Composites* **31(5)**: 285–289.

Bentz DP and Weiss WJ (2011) *Internal Curing: A 2010 State-of-the-Art Review.* U. S. Department of Commerce, Washington DC, USA, NBISTIR 7765, February. *ACI Special Publication* **278(SP)**: 13–296.

BIS (Bureau of Indian Standards) (1956) IS 516-1956 (reaffirmed 1999): Indian standard methods of tests for strength of concrete. Bureau of Indian Standards, New Delhi, India.

BIS (1970) IS 383-1970: Specification for coarse and fine aggregates from natural sources for concrete. Bureau of Indian Standards, New Delhi, India.

BIS (1987) IS 12269-1987: Specifications for 53 grade ordinary Portland cement. Bureau of Indian Standards, New Delhi, India.

Cano Barrita F de J, Bremner TW and Balcom BJ (2003) Effects of curing temperature on moisture distribution, drying and water absorption in self-compacting concrete. *Magazine of Concrete Research* **55(6)**: 517–524.

Cussion D and Hoogeveen T (2008) Internal curing of high-performance concrete with pre-soaked fine lightweight aggregate for prevention of autogenous shrinkage cracking. *Cement and Concrete Research* **38(6)**: 757–765.

Debashis D and Abhijit C (2012) A comparison of hardened properties of fly-ash-based self-compacting concrete and normally compacted concrete under different curing conditions. *Magazine of Concrete Research* **64(2)**: 129–141.

Dhir RK, Hewlett PC and Dyer TD (1994) An investigation into the feasibility of formulating self-cure concrete. *Materials and Structures* **27(174)**: 606–615.

Dhir RK, Hewlett PC and Dyer T (1995) Durability of self-cure concrete. *Cement and Concrete Research* **25(6)**: 1153–1158.

Dhir RK, Hewlett PC and Dyer TD (1998) Mechanisms of water retention in cement pastes containing a self-curing agent. *Magazine of Concrete Research* **50(1)**: 85–90.

El-Dieb AS (2007) Self-curing concrete: water retention, hydration and moisture transport. *Construction and Building Materials* **21(6)**: 1282–1287.

El-Dieb AS, El-Maaddawy TA and Mahmoud AAM (2011) Water-soluble polymers as self-curing agent in silica fume Portland cement mixes. *ACI Special Publication* **278(SP)**: 13–29.

Jensen OMP and Freiesleben H (2001) Water-entrained cement-based materials – I. Principle and theoretical background. *Cement and Concrete Research* **31(4)**: 647–654.

Jensen OMP and Freiesleben H (2002) Water-entrained cement-based materials – II. Experimental observations. *Cement and Concrete Research* **32(6)**: 973–978.

Jiajun Y, Shuguang H, Fazhou W, Yufei Z and Zhichao L (2006) Effect of pre-wetted lightweight aggregate on internal relative humidity and autogenous shrinkage of concrete. *Journal of Wuhan University of Technology – Materials Science Edition* **21(1)**: 134–137.

Jianhua J and Yingshu Y (2013) Relationship of moisture content with temperature and relative humidity in concrete. *Magazine of Concrete Research* **65(11)**: 685–692.

Liang RTY and Sun RK (2002) *Compositions and Methods for Curing Concrete*, US Patent 6, 468, 344 B1, Oct.

Lura P (2003) *Autogenous Deformation and Internal Curing of Concrete*. PhD thesis, Technical University of Delft, the Netherlands.

Machtcherine V and Reinhardt HW (2012) *Application of Superabsorbent Polymers (SAP) in Concrete Construction, State of the Art Report RILEM TC 225-SAP*. Springer, Dordrecht, the Netherlands.

Poursaee A and Hansson CM (2010) Curing time and behaviour of high-performance concrete. *Proceedings of the Institution of Civil Engineers – Construction Materials* **163(4)**: 223–230.

Wen-Chen J (2008) *Self-curing Concrete*, US Patent Application Publication, Publication No. 2008/0072799 A1, Mar.

Zhutovsky S, Kovler K and Bentur A (2002) Efficiency of lightweight aggregates for internal curing of high strength concrete to eliminate autogenous shrinkage. *Materials and Structures* **35(2)**: 97–101.

Dhir and Paine
ISBN 978-0-7277-6457-7
https://doi.org/10.1680/icetsc.64577.015
ICE Publishing: All rights reserved

Chapter 2

Mechanisms of water retention in cement pastes containing a self-curing agent

R. K. Dhir
Concrete Technology Unit, University of Dundee, Dundee, DD1 4HN, UK

P. C. Hewlett
Concrete Technology Unit, University of Dundee, Dundee, DD1 4HN, UK

T. D. Dyer
Concrete Technology Unit, University of Dundee, Dundee, DD1 4HN, UK

Self cure concrete contains a chemical agent that reduces the evaporation of water from its surface, primarily by reducing the vapour pressure at the concrete pore solution surface. The self curing agent developed at the Concrete Technology Unit, University of Dundee, also produces an alteration in cement hydration product microstructure, and it was considered that this may also contribute to the improved water retention properties. To investigate this, weight loss measurements were conducted on both self cure and ordinary pastes exposed to controlled ambient conditions, whilst thermogravimetric analysis was carried out on identical specimens. It was found that, whilst the evolution of heat of hydration renders the early stages of drying very complex, it was possible to examine the diffusion dependent stage of drying. The diffusion coefficients observed for water vapour passing through the dry region of the self cure paste surface were much lower than those observed for the control. This has been attributed to two mechanisms: the lower vapour pressure above the pore solution leading to a smaller difference across the dried portion of the paste and lower relative humidities in the cement pores, and the change in microstructure which reduces permeability.

Notation

D^*	diffusion coefficient
d	thickness of specimen
b^*	the slope of the line produced by plotting moisture loss versus time$^{1/2}$
M_{ew}	the mass of evaporable water

Introduction

The necessity for the presence of sufficient water in concrete to allow the binder constituents to react to an adequate extent is of vital importance in concrete construction. Evaporation of water from concrete is therefore undesirable, and its effect on the development of microstructure is most noticeable at the surface,[1-3] leading to poor concrete durability. As a result, curing must be conducted in most circumstances to reduce evaporation. However, in many applications, good curing is not always practical, and work has been conducted at these laboratories into developing an agent that, when added to concrete that is subsequently cured in air, leads to the retention of more water when compared to an identically cured control. The main way in which this improved water retention is effected is by the use of water soluble polymers that take part in hydrogen bonding with water molecules. This produces a 'negative deviation' from Raoult's law (the relationship that describes the partial vapour pressure of a component of an ideal solution above its surface),[4] and hence reduces the rate at which water evaporates.

During this work, it became apparent that the chemical selected as being most suitable for use as a self-cure agent (a water-soluble, polymeric glycol) produced concrete water retention far superior to that observed when the chemical is present in water alone. Modifications in the micro structure of cement hydration products (in particular calcium hydroxide) caused by the addition of this chemical have been observed previously using X ray diffraction and scanning electron microscopy techniques.[5] When viewed through the scanning electron microscope, self-cure cement pastes display numerous, extremely thin crystals of calcium hydroxide.[5] It is probable that the presence of these crystals refines the pore structure of the paste,[6] since these modifications appear to strongly influence concrete permeability.

Evidence for the micro structural changes influencing permeability can be seen in Fig. 1, which shows intrinsic permeabilities deduced from air permeability measurements[7] made on cores taken from 100 mm concrete cubes. Results from concrete made from three different binder types are shown (100% PC, 60% PC/40% GGBS and 70% PC/30% PFA), which were cured

Fig. 1 Effect of curing technique on intrinsic permeabilities of cores taken from the surfaces of three types of concrete (PC, Portland cement; GGBS, ground granulated blastfurnaceslag; PFA, pulverized fuel ash)

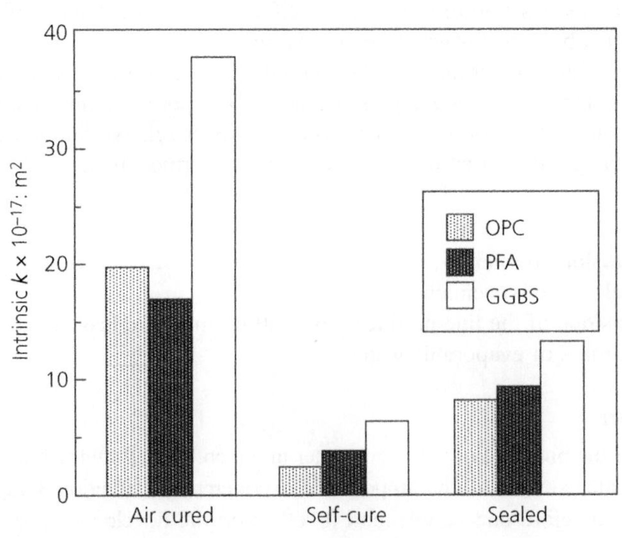

for 24 h under damp hessian before being transferred to a 20°C/60% relative humidity curing environment for 28 days. The self-cure specimens, which contained the admixture in the mix water at a concentration of 0·1 M, were cured alongside two sets of controls containing no admixture – air-cured controls which were exposed, and sealed controls which were wrapped in a waterproof film. It is clear that for all three binder types the self-cure specimens are less permeable than both the air and sealed controls. This suggests that the modification in microstructure improves self-cure concrete performance, since it would be expected that the self-cure specimens would possess permeabilities between those of the two types of control. From these permeability results, it was considered possible that this alteration in microstructure was contributing to the water retaining capabilities of the cement matrix.

This paper reports the results of an investigation into the way in which moisture is lost from self-cure and normal cement pastes.

Drying behaviour of porous solids

Drying of a porous solid occurs by the evaporation of the fluid it contains. The driving force behind the evaporation of a fluid is the difference in chemical potential between the evaporable liquid and its vapour in the environment.[4] Evaporation therefore occurs to reduce this difference, and reduce the Gibb's free energy. In a closed system, evaporation proceeds until the change in the Gibb's free energy is zero, and hence the vapour pressure above the liquid is sufficient for the two phases to be in equilibrium. However, when the system is open, as in the case of a porous solid exposed to the atmosphere, this equilibrium is never achieved, since the vapour pressure above the liquid is constantly reduced as vapour disperses. Therefore, for an open system, evaporation will continue indefinitely as long as there is sufficient fluid.

The drying behaviour of most saturated porous solids can be divided into three stages, identified by plotting the rate of drying against moisture content.[8] A typical plot is shown in schematic form in Fig. 2.

Fig. 2 Typical plot of drying rate versus moisture content

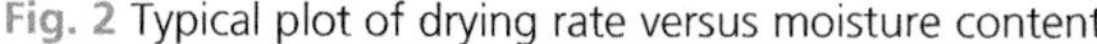
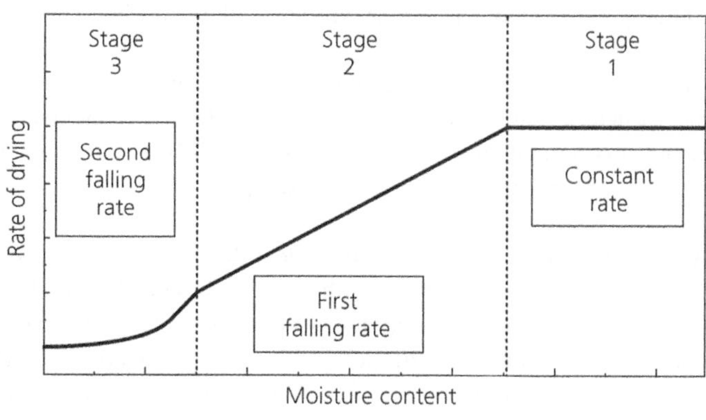

The first stage of drying, known as the constant rate period, occurs by evaporation of liquid at the material's surface. The fluid at the surface is replenished by the flow of more liquid from within the solid. The rate of drying is constant, since the situation is much the same as for a free liquid.

The second drying stage is referred to as the first falling rate period, and initiates when the saturation of liquid in the surface region drops to a level insufficient for the transport of liquid at a rate to match that of evaporation. As a result the saturation at the surface drops more rapidly.

The final stage of drying, the second falling rate period, begins as the saturation at the solid's surface reaches its residual level and flow of liquid ceases completely in this region. This means that as further evaporation occurs a dry front develops at the surface which gradually penetrates further into the solid. During the second falling rate period, the rate of moisture loss becomes dependent on the rate at which the liquid vapour can diffuse through the dry front,[9] which is controlled by the difference in vapour pressure across the dry front and the micro structure of this region. The diffusion coefficient (D^*) can be calculated using the following equation:[10]

$$D^* \times 10^6 = 218 \cdot 7(db^*/M_{ew})\,\mathrm{m}^2/\mathrm{s}$$

where d is the thickness of specimen, b^* is the slope of the line produced by plotting moisture loss versus time$^{1/2}$ and M_{ew} is the mass of evaporable water.

The drying of fresh cement paste and concrete

Whilst freshly mixed cement pastes and concrete possess the correct physical attributes to suggest that they might dry in the manner described above, three characteristics of Portland cement mean that the situation is more complex. One factor is the evolution of heat during cement hydration: the increase in temperature in the hours subsequent to mixing leads to higher rates of evaporation.[11] Also, the development of microstructure as hydration proceeds will influence the drying behaviour, particularly during the final period.[12] Finally, another important factor is the uptake of free water during hydration: simply measuring specimen weight loss and subtracting this from the original water content will not give the true moisture content, since a large proportion of this water will have taken part in cement reactions.[13]

When investigating the water retention mechanisms in self-cure concrete it was recognized that the technique for determining moisture content required some refinement. The technique selected to carry out this refinement was thermogravimetry (TG), since this allows a direct measurement of chemically bound water in cement samples.[14]

As already discussed, the rate at which moisture is lost during the final stage of drying in a porous medium is governed by the rate at which the vapour diffuses through the dry front. The diffusion coefficient of water vapour in hardened concrete and cement paste, however, is dependent on the relative humidity in the pores. There exists a critical relative humidity (around 75%) above which the diffusion coefficient is relatively high and below which it is low.[15] It is believed that the low coefficient values at low humidities are a result of the movement of water along solid surfaces becoming the predominant diffusion mechanism. As a result, the diffusion coefficient of cement paste and concrete is dependent on the concentration of evaporable water in a specimen.[12]

Experimental procedures

Two cement paste mixes were investigated, one containing the self-cure agent at a concentration of 0.1 M in the mix water, and a control. A Portland cement and distilled, deionized water were used throughout. Sixteen polypropylene beakers with an internal diameter of 56 mm and a depth of 70 mm were filled with 100 g of cement. The beakers were then transferred to a 20°C/60% relative humidity environment. Fifty grams of the self-cure agent solution was added to eight of the beakers, whilst the remaining beakers were filled with the same amount of distilled water. The pastes were stirred by hand for 2 min. The weight of one of the self-cure and one of the control beakers was measured, and these beakers were thence onwards used for weight-loss measurements, which were carried out at frequent intervals for the first 24 h and continued for 28 days. In addition, cement paste of the same water/cement (w/c) ratio was cast in sealed containers of similar dimensions and kept in the same environment.

At ages of 3, 6, 12 and 24 h, and 3, 7, 14 and 28 days, one of each of the paste samples was removed and ground with acetone. They were then washed with a further quantity of acetone and dried at 50°C in a vacuum oven for 2 h. Ten milligrams of each sample was then analysed using TG. TG was conducted using a temperature ramp of 10°C/min up to 1000°C in continually refreshed nitrogen.

Bleeding of both self-cure and control pastes was measured. This was done by filling 25 mm diameter polyethylene cylinders with the pastes up to a depth of 110 mm. The depth of bleed water was then measured at 10 min intervals for 2 h. Measurements were made to the nearest 0.5 mm at four points around the circumference of the cylinders and a mean depth calculated.

Results and discussion

Figure 3(a) shows weight loss with time for the two pastes. It is clear that throughout the 28 day test period the loss of moisture by evaporation from the self-cure paste is less than that of the control. When these data are used to create a plot of rate of moisture loss versus moisture content (Fig. 3(b)) the results from both pastes bear a poor resemblance to the typical behaviour of drying porous media. As discussed previously this can be attributed to the evolution of heat and the chemical binding of water during hydration.

The effect of the evolution of heat can be seen by referring to calorimetry curves obtained from identical paste mixes, shown in Fig. 4(a). The heat evolution peaks for both pastes can be matched to distinct shoulders in plots of the rate of moisture loss with time (Fig. 4(b)). The self-cure agent has a slight retarding effect on the production of cement hydration products and hence the second boosting of evaporation rate occurs later.

Figure 5 shows that the admixture has the effect of reducing bleeding. This is likely to reduce the amount of moisture loss during stage 1. However, since the rate of evaporation from the bleed water at the surface of the drying self-cure pastes will evaporate at a lower rate, the overall significance of reduced bleeding is not great.

Fig. 3 (a) Weight loss from paste specimens over a 28 day period and (b) drying plots constructed from these data

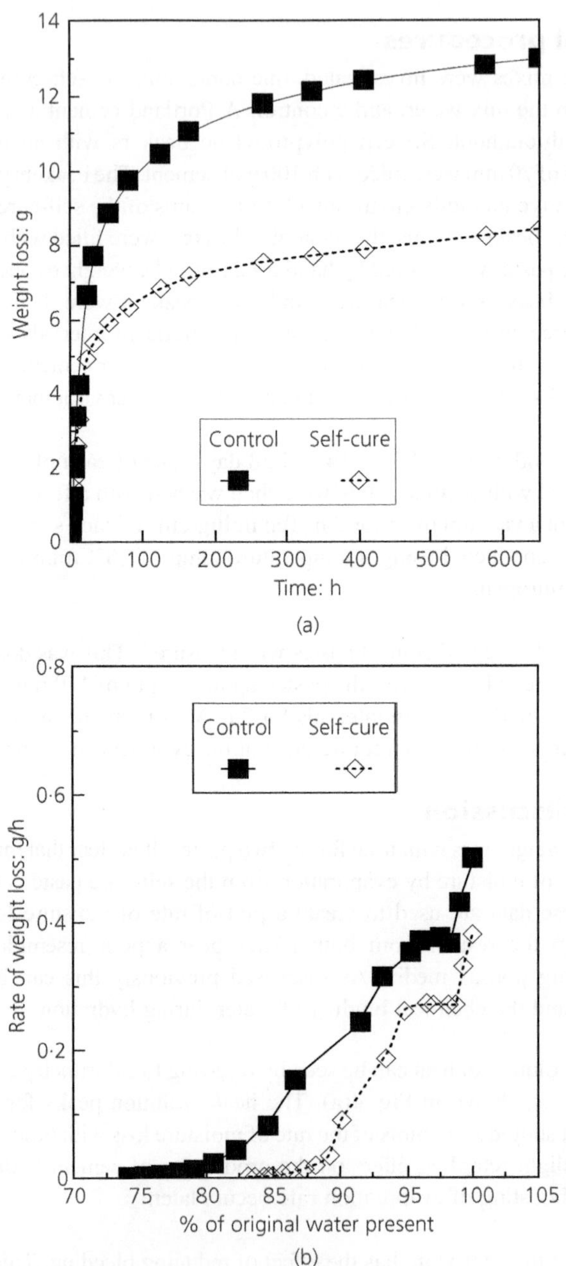

Fig. 4 (a) Calorimetry curves obtained from the two paste mixes, and (b) the rate of weight loss from paste samples with time

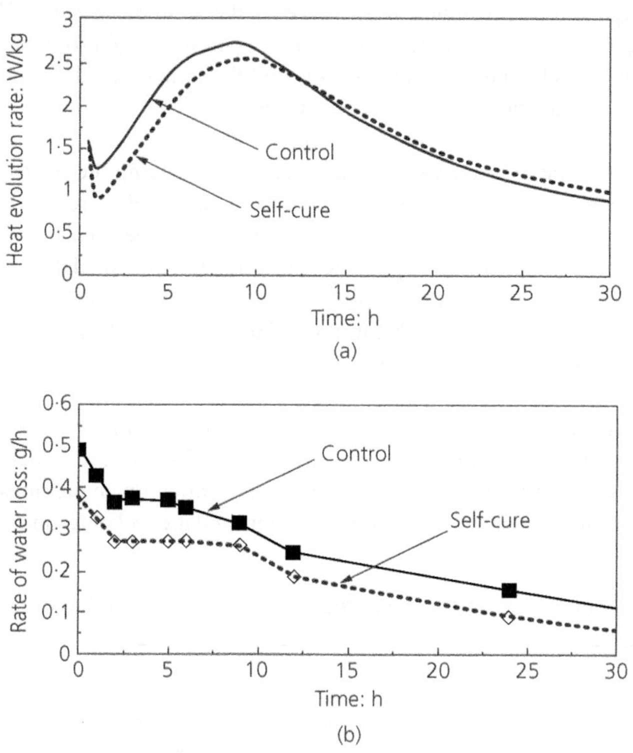

Fig. 5 Bleed characteristics

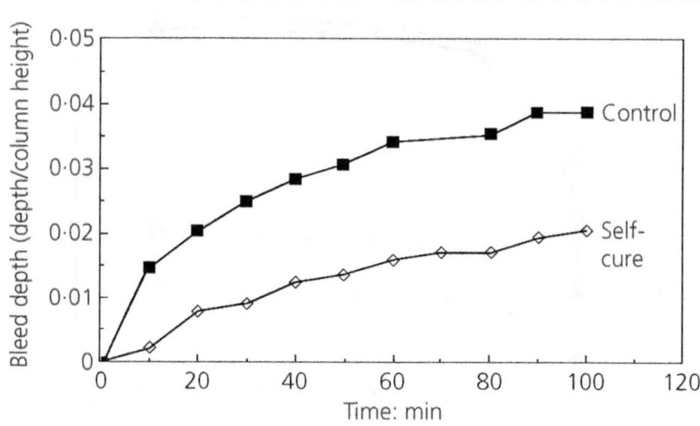

Having conducted TG on the dried cement paste samples, the total quantity of chemically bound water was found by obtaining the total weight loss at 1000°C and then subtracting the quantity of CO_2 evolved by the thermal decomposition of calcite (this was found to be small in all cases, although somewhat higher in the case of the self-cure pastes). These results are shown in Fig. 6, along with evaporated water values from the weight-loss measurements. The self-cure paste displays bound water levels only slightly higher than the control, the result of higher degrees of carbonation in the self-cure specimens.

Figure 7 shows evaporable water values calculated by subtracting the bound and evaporated water values from the original water content. Included with the self-cure and control results are the evaporable water values obtained for the sealed paste specimens calculated by subtracting the bound water values from the original water content. The calculated evaporable water values were used to construct new drying plots (Fig. 8). Whilst heat evolution renders the early stages of drying fairly complex, the beginning of the second falling rate period, at around 24 hours for both specimens, is clearly defined.

It is during the second falling rate period that the microstructure of the cement paste becomes most influential on the drying process, since it will control the rate at which vapour diffuses through the developing dry front. Using the diffusion equation given previously, the diffusion coefficient was calculated using data between 24 hours and the end of testing. In both cases the

Fig. 6 Bound water in cement pastes (measured using TG) and evaporated water

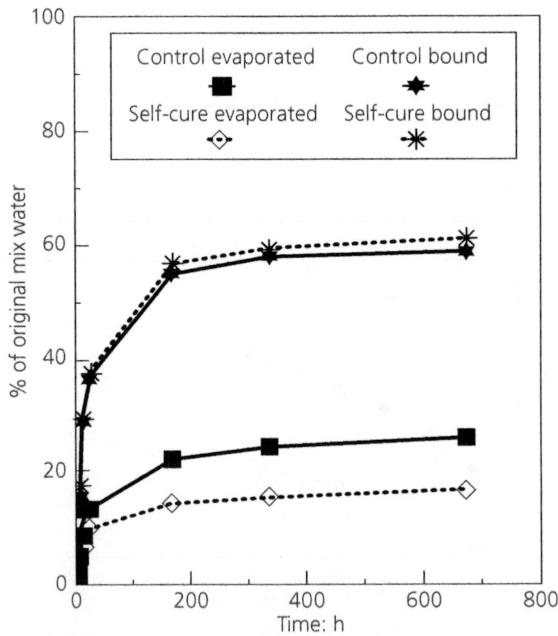

Fig. 7 Total evaporable water in cement paste specimens calculated using bound and evaporated water data

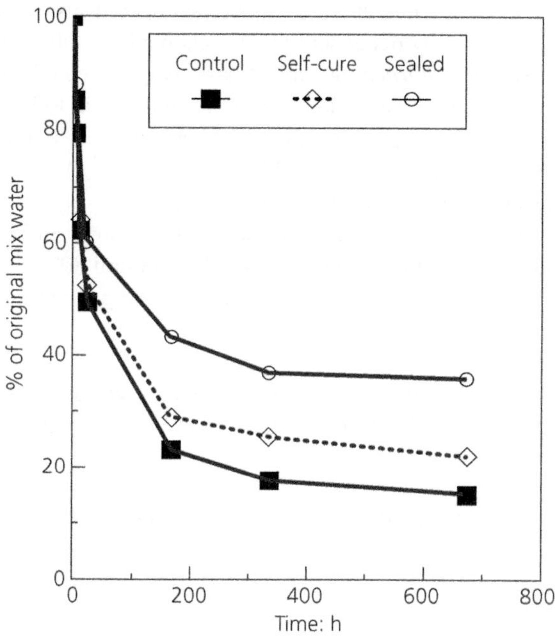

Fig. 8 Paste drying plots constructed using the calculated evaporable water values

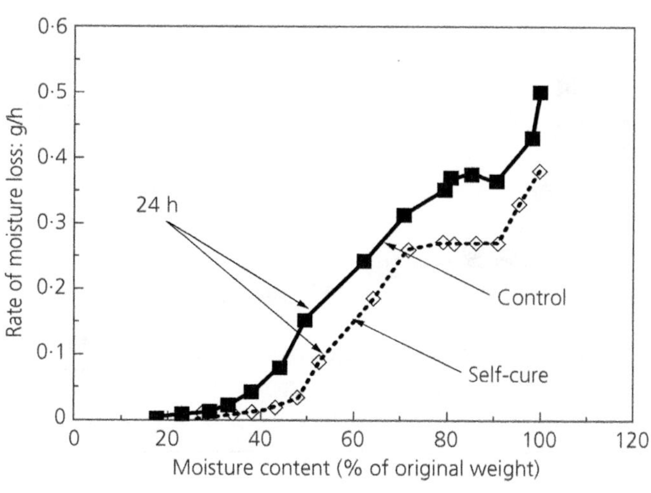

diffusion coefficient decreases as the dry front penetrates into the samples. This can be attributed to three factors.

(*a*) As discussed previously, the rate at which water vapour diffuses through hardened cement and concrete is highly dependent on the relative humidity in the pores. Whilst water vapour diffusion in a *hardening* cement paste may not be effected in the same manner, it is probable that, at least after setting, this will begin to become a factor. As the dry front penetrates deeper, the effect will become more important as more of the pores' humidities fall below the critical level.[15]

(*b*) The cement grains at the surface are less densely packed.

(*c*) As the dry front penetrates below the paste surface, hydration will be arrested when the relative humidity at a particular depth falls below around 80%.[16] The surface region will be the first to stop hydrating, and hence will have the poorest microstructural qualities.

Figure 9 shows the calculated diffusion coefficient values versus time. Whilst there will always be a difference between the diffusion coefficient values obtained for the two pastes

Fig. 9 Diffusion coefficient values calculated for drying pastes as a function of time

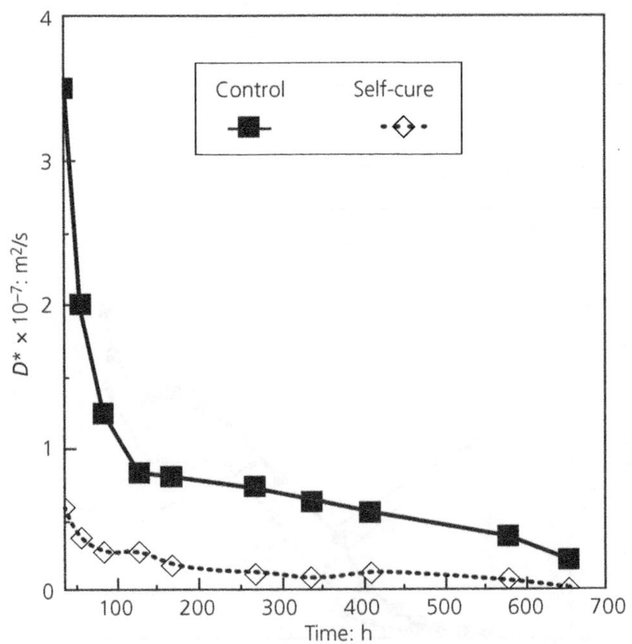

(due to the lower vapour pressure above the self-cure solution, and hence a lower difference between this pressure and the atmospheric vapour pressure), the diffusion coefficient values obtained for the self-cure paste are much lower than for the control. This would suggest that one or both of the following are occurring:

(*a*) The self-cure agent is reducing the vapour pressure at the pore water surface sufficiently to bring the diffusion coefficient down below the 75% critical level over a greater portion of the dry front.
(*b*) The microstructure at the surface of the self-cure paste is less permeable to water vapour than that of the control.

Figure 10 shows the calculated diffusion coefficients as a function of evaporable water concentration. Again, the self-cure paste displays considerably lower diffusion coefficient values than the control. The large differences between these values at early ages suggests that both of the mechanisms proposed above are aiding in water retention, at least early on in the second falling rate period.

Fig. 10 Diffusion coefficient values calculated for drying pastes as a function of moisture content

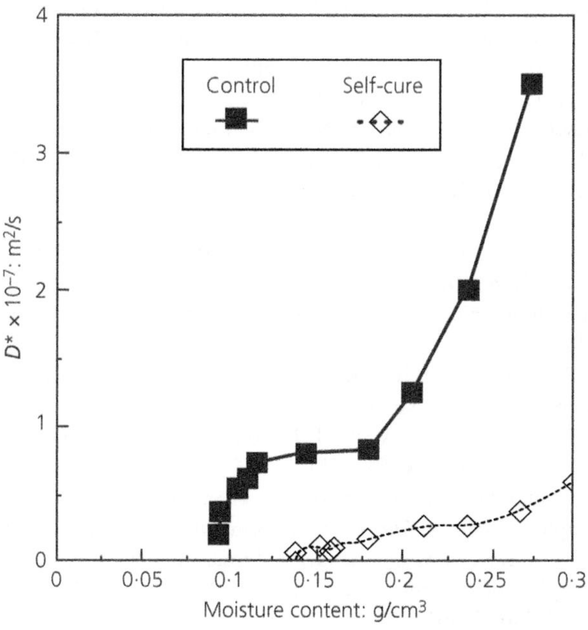

Conclusions

Three mechanisms by which water retention is improved in self-cure cement paste have been identified:

- *Lowering of vapour pressure*, which reduces the rate of evaporation throughout the drying process, and also is potentially a factor in reducing the rate of water vapour diffusion during the third stage of drying.
- *Reduction of bleeding*, which reduces the amount of water lost during the first stage of drying.
- *Reduction of permeability*, by modification of hydration product micro structure, which contributes to the reduction in the rate of water vapour diffusion during the final stage of drying.

REFERENCES

1. Dhir R. K., Hewlett P. C. and Chan Y. N. Near surface characteristics and durability of concrete: an initial appraisal. *Magazine of Concrete Research*, 1986, **38**, No. 134, 54–56.
2. Dhir R. K., Hewlett P. C. and Chan Y. N. Near surface characteristics of concrete: intrinsic permeability. *Magazine of Concrete Research*, 1989, **41**, No. 147, 87–97.
3. Dhir R. K., Hewlett P. C. and Chan Y. N. Near-surface characteristics of concrete abrasion resistance. *Materials and Structures*, 1991, **24**, 122–128.
4. Atkins P. W. *Physical Chemistry*, 5th edn. Oxford University Press, Oxford, 1994.
5. Dhir R. K., Hewlett P. C., Lota J. S. and Dyer T. D. An investigation into the feasibility of formulating 'self-cure' concrete. *Materials and Structures*, 1994, **27**, 606–615.
6. Meng B. Calculation of moisture transport coefficients on the basis of relevant pore structure parameters. *Materials and Structures*, 1994, **27**, 125–134.
7. Dhir R. K., Hewlett P. C. and Chan Y. N. Near-surface characteristics of concrete – intrinsic permeability. *Magazine of Concrete Research*, 1989, **41**, No. 147, 87–97.
8. Dullien F. A. *Porous Media: Fluid Transport and Pore Structure*. Academic Press, London, 1979.
9. Heller J. P. The drying through the top surface of a vertical porous column. *Soil Science Society of America Proceedings*, 1968, **32**, 778–786.
10. Mills R. H. Mass transfer of water vapour through concrete. *Cement and Concrete Research*, 1985, **15**, 74–82.
11. Wang J., Dhir R. K. and Levitt M. Membrane curing of concrete: moisture loss. *Cement and Concrete Research*, 1994, **24**, No. 8, 1463–1474.
12. Lowe I. R. G., Hughes B. P. and Walker J. The diffusion of water in concrete at 30°C. *Cement and Concrete Research*, 1971, **1**, 547–557.
13. Bažant Z. P. Constitutive equation for concrete creep and shrinkage based on the thermodynamics of multiphase systems. *Materials and Structures*, 1970, **3**, No. 13, 3–36.
14. Taylor H. F. W. *Cement Chemistry*. Academic Press, London, 1990.
15. Bažant Z. P. Drying of concrete as a non-linear diffusion problem. *Cement and Concrete Research*, 1971, **1**, No. 5, 461–473.
16. Patel R. G., Killoh D. C., Parrott, L. J. and Gutteridge W. A. Influencing of curing at different relative humidities upon compound reactions and porosity in Portland cement paste. *Materials and Structures*, 1988, **21**, 192–197.

Self-healing

Dhir and Paine
ISBN 978-0-7277-6457-7
https://doi.org/10.1680/icetsc.64577.029
ICE Publishing: All rights reserved

Chapter 3

Self-healing cementitious materials: a review of recent work

Christopher Joseph MEng, PhD
Research assistant, Cardiff School of Engineering, Cardiff University, UK

Diane Gardner MEng, PhD
Lecturer, Cardiff School of Engineering, Cardiff University, UK

Tony Jefferson MSc, PhD, CEng, MIStructE, MICE
Reader, Cardiff School of Engineering, Cardiff University, UK

Ben Isaacs MEng, PhD
PhD student, Cardiff School of Engineering, Cardiff University, UK

Bob Lark PhD, CEng, MICE
Professor, Cardiff School of Engineering, Cardiff University, UK

Historically construction materials have been designed to meet a fixed specification and material degradation has been viewed as inevitable and mitigated for through expensive maintenance regimes. Material scientists have recently begun developing materials which have the ability to adapt and respond to their environment, drawing on their knowledge and familiarity of biological systems. This fundamental change in material design philosophy has resulted in the creation of a whole host of 'smart' materials, including self-healing materials. The development of self-healing materials is reviewed in this paper, together with definitions of common terminology. A brief summary of the construction industry is given, together with a synopsis of the main issues of durability relating specifically to cementitious materials. Specific focus is then given to both autogenic (natural) and autonomic (manufactured) healing processes within cementitious materials. The paper concludes with a summary of self-healing materials, an overview of their potential use within the construction sector, and recommendations to this sector for future uptake of these new and innovative materials.

1.   Introduction

1.1   Background

The construction industry accounts for approximately 10% of the UK gross domestic product (Corporatewatch, 2004). The industry, in its widest sense, is responsible for all aspects of the built environment; as such, the materials used for construction are of utmost importance, not only in terms of initial physical and aesthetic properties but also in terms of durability and longevity.

Historically construction materials have been designed to meet a fixed predefined specification, and material degradation has been viewed as an inevitable process which is commonly addressed and managed through expensive maintenance regimes. Recently, however, material scientists, often gaining their inspiration from biological systems, have begun to challenge this view, and have been developing materials which have the ability to adapt and respond to their environment. This fundamental change in material design philosophy has resulted in the creation of a whole host of 'smart' materials, including self-healing materials.

Evidence of this new and innovative approach to material design was apparent at the first international conference on self-healing materials which was held in the Netherlands during April 2007. This conference has also been succeeded by the first two books devoted solely to self-healing materials (Ghosh, 2009; Van der Zwagg, 2007), and in 2005 a Rilem committee was established charged with investigating self-healing phenomena in cement-based materials (Schlangen, 2005). The growing interest in the arena of self-healing materials was demonstrated by the size and success of the second international conference on self-healing materials, held in Chicago in July 2009.

The self-healing abilities of a wide range of materials have been explored to date, including metals, polymers, ceramics/concretes and coatings. The motivation for developing these materials is to improve their long-term durability and service life, thereby reducing whole lifetime costs, both in terms of economic and environmental impact. Much of the work on self-healing materials remains in the early stages of development; however, some of these technologies have already made the transition into the commercial sector, such as scratch-healing paints for the automotive industry (Nissan Motor Co., 2005).

1.2   Definition of terms

There are many terms used within the literature to describe the development of new composite materials designed to possess some 'higher-order' function. The self-healing materials discussed in this review fall under the category of smart structures, as indicated in Figure 1. Smart structures are engineered composites of conventional materials, which exhibit sensing and actuation properties, due to the properties of the individual components. Many self-healing

Figure 1 Schematic representation of self-healing terms

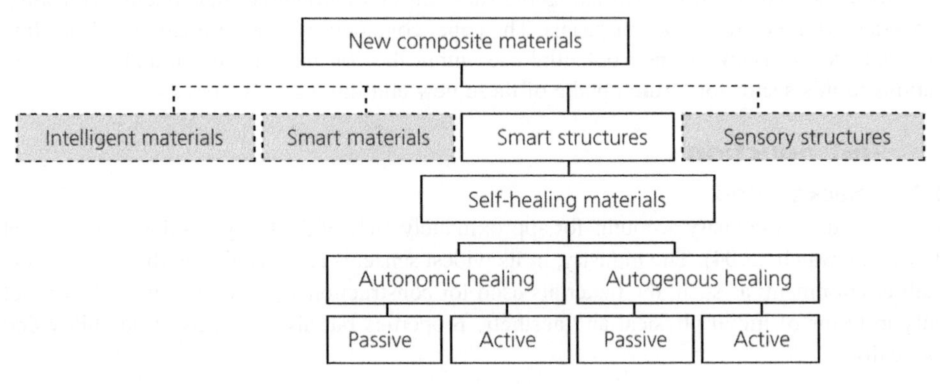

materials fall into the category of smart structures, since they contain encapsulated healing agents which are released when damage occurs, thereby 'healing' the 'injury', and increasing the materials' functional life.

A composite material which exhibits self-healing capabilities due to the release of encapsulated resins or glues, as a result of cracking from the onset of damage, is categorised as having autonomic healing properties: the composite has been 'manufactured' to exhibit healing behaviour.

If the healing properties of a material are generic to that material, then the material could potentially be classed as a smart material, and the healing process is termed autogenic healing: the material exhibits a 'natural' healing ability. Cementitious materials have this innate ability to self-repair, since rehydration of a concrete specimen in water can serve to kick-start the hydration process, when the water reacts with pockets of unhydrated cement in the matrix.

Smart self-healing structures may also be classified depending on the passive or active nature of their healing abilities. A passive mode smart structure has the ability to react to an external stimulus without the need for human intervention and hence benefits from omitting the need for human inspection, repair and maintenance. An active mode smart structure requires human intervention in order to complete the healing process. Both systems have been tested, in respect to concrete, by Dry (1994).

2.   The driving force behind self-healing materials

Construction materials form a large and important sector within the construction industry. Despite an increasing trend towards importation of building materials, over £9 billion (Office for National Statistics, 2008) is generated from the export market. The potential benefit to the UK economy from the development of new and innovative construction materials, such as self-healing concrete, is therefore significant.

The durability and longevity of the materials used is of paramount importance, particularly when one considers that a typical design life for a civil engineering structure is 50 years or more. It can be seen from Figure 2 that repair and maintenance costs account for approximately 45% of the total expenditure on construction. Given the very wide use of concrete within the sector then it is fair to assume that a substantial proportion of these maintenance costs may be attributed directly to poor durability of concrete structures. It should also be noted that the maintenance costs over the last decade are likely to reflect the legacy of poor construction methods and quality control in the 1960s and 1970s.

Cracking of concrete, especially microcracking, is extremely common, and can be caused by a number of factors including structural loading, plastic shrinkage, drying shrinkage, and thermal effects. Cracking leads to an increase in the number of pathways open to the ingress of saline water, acid rain, and carbon dioxide. Depending on the type of concrete, the environmental conditions and the chemical composition of the infiltrating fluid, various degradation processes can occur, causing further cracking and substantially compromising the durability of the structure. In addition to reducing the structural integrity of the concrete, cracking is aesthetically unsightly and also increases the porosity of the material, thus potentially compromising the watertightness of the concrete from which the structure is formed. This is a very important requirement for water-retaining structures

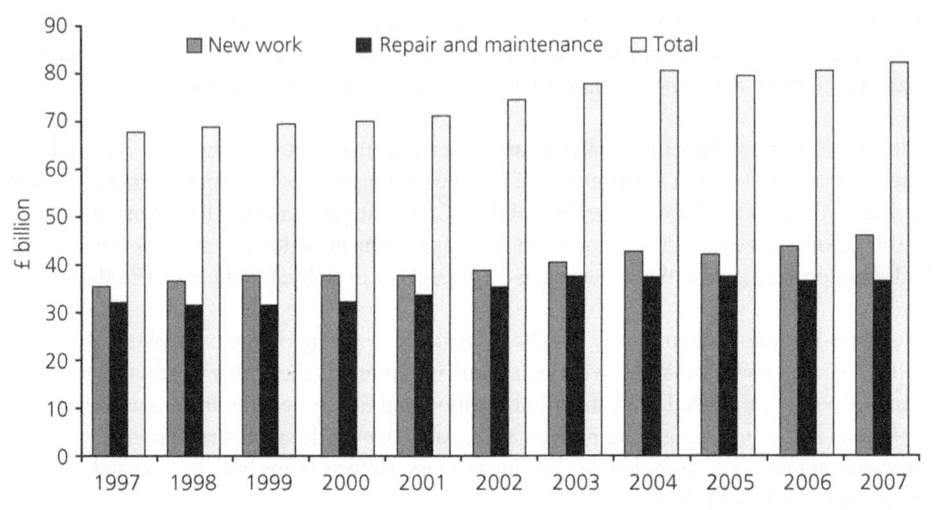

Figure 2 Billions of pounds spent on repair and maintenance and new construction work in Great Britain between 1997 and 2007 (Constant 2000 prices, seasonally adjusted) (Office for National Statistics, 2008)

such as dams and storage tanks but also for other critical housing facilities such as nuclear reactors and disposal facilities for radioactive waste.

Conventional methods for improving the durability of concrete involve the reduction of cracking through the use of fibre-reinforcing, or the reduction in permeability through the use of waterproofing agents, sealants, or the creation of denser microstructures with reduced porosity.

In the context of concrete durability, in particular, the ability to self-repair microcracks through the release of an encapsulated healing agent or through accelerating the natural autogenous healing process, self-healing will have a profound effect on the longevity of concrete structures.

3.   Overview of self-healing materials

This review investigates the current situation in relation to the development of self-healing materials within the construction industry. Particular focus has been placed on the process of autonomic and autogenic healing and their associated technology. The application of this technology to cementitious materials has been considered. However, initial reference is made to polymeric materials because a far larger body of research has been completed on the autonomic healing of polymeric materials, and therefore this offers potential opportunity for technology transfer between the different material types.

3.1   Autonomic healing in polymers

Autonomic-healing polymers have been the subject of much attention since their initial creation in the material laboratories of the University of Illinois at Urbana-Champaign. Their recent development is widely credited to the team at this institution led by Professor White, and significant literature has been published in this area (Brown *et al.*, 2002, 2003a, 2003b, 2004, 2005a, 2005b; Kessler *et al.*, 2003; White *et al.*, 2001). White *et al.* (2001) originally recognised that thermosetting polymers are susceptible to damage in the form of cracks, which form deep within the structure where detection is difficult and repair is almost impossible. Cracking leads to mechanical degradation of fibre-reinforced polymer composites, and in microelectronic polymer components it can also lead to electrical failure. The self-healing concept in polymers was introduced by Kessler *et al.* (2003). The self-healing concept concerns a microcapsule embedded in a structural matrix containing a catalyst. The first stage is characterised by the propagation of a crack through the matrix due to the occurrence of damage, such as fatigue loading. The second stage is when a microcapsule ruptures, thereby releasing the healing agent into the crack plane through capillary action. The third, and final, stage is the contact between the healing agent and the catalyst, resulting in polymerisation and 'healing' via bonding of the crack faces. A schematic representation of this can be found in Kessler *et al.* (2003).

3.2   Autonomic healing in cementitious materials

The autonomic healing of cementitious composites has received significantly less attention than its polymer counterpart, and the research efforts to date have been far more piecewise and sporadic. Nevertheless, several authors (Dry, 1994, 1996a, 1996b, 1996c, 1996d, 2000; Dry and Corsaw, 2003; Dry and Unzicker, 1998; Li *et al.*, 1998; Mihashi *et al.*, 2000) have completed studies in this area.

The concept of autonomic healing in concrete was originally proposed for cementitious materials by Dry (1994). This concept, as illustrated in Figure 3, is similar to that described previously for

Figure 3 Autonomic healing concept in cementitious materials (Joseph *et al.*, 2010)

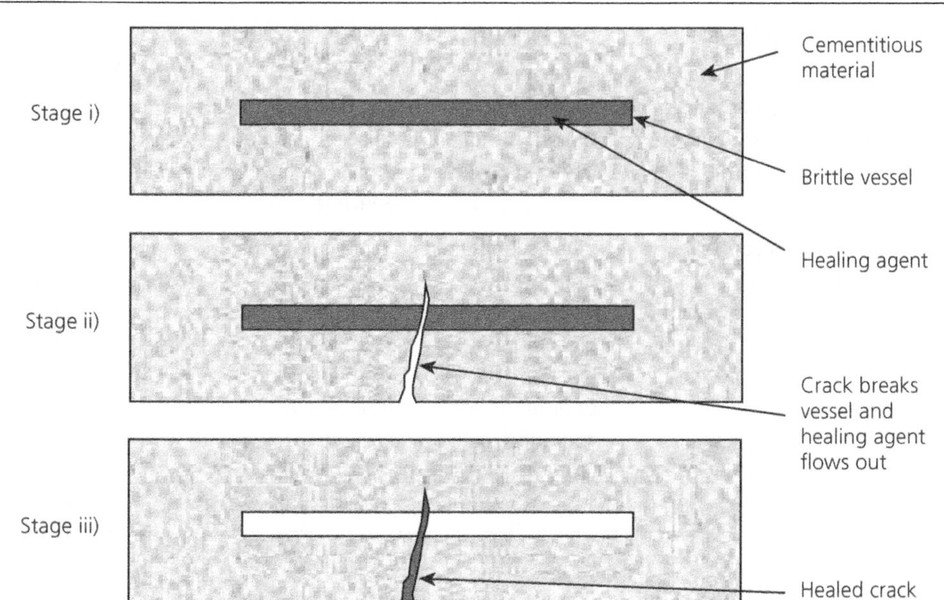

polymers. Due to the vastly different structure of concrete compared to plastic and the infancy of this research area, the sophistication of the encapsulating methods and the encapsulated materials are somewhat different, although the underlying philosophy is the same.

3.2.1 Healing agents

Various healing agents have been proposed in studies undertaken on the self-healing of cementitious materials. These healing agents have generally been 'off the shelf' agents and as such are reasonably low cost and readily available, essential attributes a healing agent should possess in order to be proposed for application to a large bulk material, such as concrete.

The main healing materials that have been proposed to date are epoxy resins, cyanoacrylates (superglues), and alkali–silica solutions. It is obvious that the effectiveness of the healing process in not only dependent on the crack width, which dictates the capillary forces, but also on the viscosity of the repair agent; the lower the viscosity the larger the potential repair area. Another prerequisite for the agent is that it must form a sufficiently strong bond between the surfaces of the crack in order to prevent the reopening of the crack, and thus force other new cracks to open, hence increasing the total fracture energy that is required to break the specimen. The main types of healing agent are summarised in Table 1.

Table 1 Summary of healing agents used in autonomic healing of cementitious materials

	Healing agent			
	Epoxy resin 1 part	Epoxy resin 2 part	Cyanoacrylate	Alkali–silica solution
Viscosity (centipoises)	200 to 500	200 to 500	< 10 (can heal cracks 100 µm)	Unknown
Method of activation	Heat	Hardener and resin	Moisture	Oxygen (causes rehydration)
Main advantage	Good thermal/ moisture and light resistance	Good thermal/ moisture and light resistance	Rapid curing due to acidity of glue and alkalinity of concrete	Low compatability problems
Main disadvantage	Sensor/heat source required for activation	Small chance that both activation components are present at crack location and poor mixing of components	Rapid curing may lead to poor dispersion of glue Potential compatability problems with concrete	For optimal results requires crack faces to be completely closed

Water has a viscosity of 1 centipoise (1 m Pa s), milk is 3 centipoise, and grade 10 light oil is 85–140 centipoises.

3.2.2   Encapsulation techniques

Various encapsulation techniques have been proposed in the literature. Mihashi *et al.* (2000) discuss two encapsulation techniques – namely a microcapsule enclosing repairing agent mixed in concrete; and a continuous glass supply pipe enclosing repairing agent, embedded in concrete – while Li *et al.* (1998) used cyanoacrylate enclosed in capillary tubes, sealed with silicon.

The shape of the embedded capsule is a factor which should be considered. A spherically shaped capsule will provide a more controlled and enhanced release of the healing agent upon breakage, and will also reduce the stress concentrations around the void left from the empty capsule. A tubular capsule, however, will cover a larger internal area of influence on the concrete for the same volume of healing agent (higher surface area to volume ratio). The release of the healing agent upon cracking however will be of inferior quality as localised and multiple cracking may occur inhibiting the effective distribution of the healing agent.

3.2.3   Mechanical response of self-healed specimens

Li *et al.* (1998) undertook their self-healing experiments within the matrix of an engineered cementitious composite (ECC), a fibre-reinforced cementitious composite that exhibits significant tensile strain-hardening characteristics due to the inclusion of high-modulus polyethylene fibres. These fibres serve to constrain the crack width opening of the matrix, thus giving a large, controlled, ductile mechanical response.

The authors investigated the mechanical response of a number of three-point and four-point bend beams, with and without healing agent encapsulated in discrete capillary tubes, subjected to damage via mechanical loading. Evidence of the healing mechanism was evaluated through studying the beam stiffness on repeated flexural loading. Three load–reload cycles were performed, each with a delay of 5 min between cycles to allow the healing agent to set. Displacement control was used to achieve a central deflection of 1·5 mm, 2·5 mm and then failure of the specimen.

The results indicated that whereas over half of the specimens showed higher regained stiffness on first reloading than the initial stiffness, the remainder did not due to hardening of the sealing agent in the capillary tube prior to testing. Furthermore, the regaining of stiffness in the second reloading cycle is less significant. As suggested by the authors, this is probably due to exhaustion of available sealing agent in the first damage/healing cycle.

Mihashi *et al.* (2000) also performed a study on the mechanical response of self-healed three-point bend specimens. However, unlike Li *et al.* (1998) the authors examined the effect of self-healing on normal notched concrete beams, subjected to the continuous supply of healing agent through glass tubes. The healing properties of both two-part epoxy resin and alkali–silica solutions were investigated, and compared to control specimens, and specimens where the healing agent had been actively injected into surface cracks. Both the diluted and undiluted alkali–silica solutions showed increased strength recovery. As expected, the manual injection of the epoxy resin produced very good strength recovery results; however, no significant effect was measured when this healing agent was encapsulated, due to insufficient mixing of the two components of the resin. Moreover, despite the continuous supply of healing agents in these experiments the authors found self-healing to be significant when the crack width is less than approximately 0·5 mm. For damage greater than this level the size of the cracks are believed to be too large to self-heal effectively.

More recently, researchers at Cardiff University have undertaken a substantial parametric study on adhesive-based self-healing of mortar beams, with the long-term aim of developing a material model for this new composite material (Joseph, 2008; Joseph *et al.*, 2010). Preliminary work undertaken by the authors revealed that: (*a*) a low-viscosity, single-agent, cyanoacrylate adhesive offered the greatest chance of flowing into the crack plane prior to curing, and (*b*) a fully encapsulated delivery system (adhesive sealed within short lengths of hollow capillary tubes) did not allow the adhesive to readily flow out following cracking (Joseph, 2008).

The experimental set-up consisted of four hollow capillary tubes open to the atmosphere, filled with a low viscosity cyanoacrylate, encased within mortar beams, as illustrated in Figure 4. Each set of experiments consisted of two control beams and four self-healing (SH) beams. The capillary tubes in the SH beams were filled with cyanoacrylate and the control beams were filled with ink, or left empty, prior to three-point bend testing (Figure 5). All tests were conducted in two stages, whereby a predefined amount of damage (crack mouth opening displacement (CMOD)) was caused during the first three-point bend test, followed by unloading and then reloading to failure after a curing period of 24 h.

Figure 4 Testing arrangement for self-healing experiments on notched beams

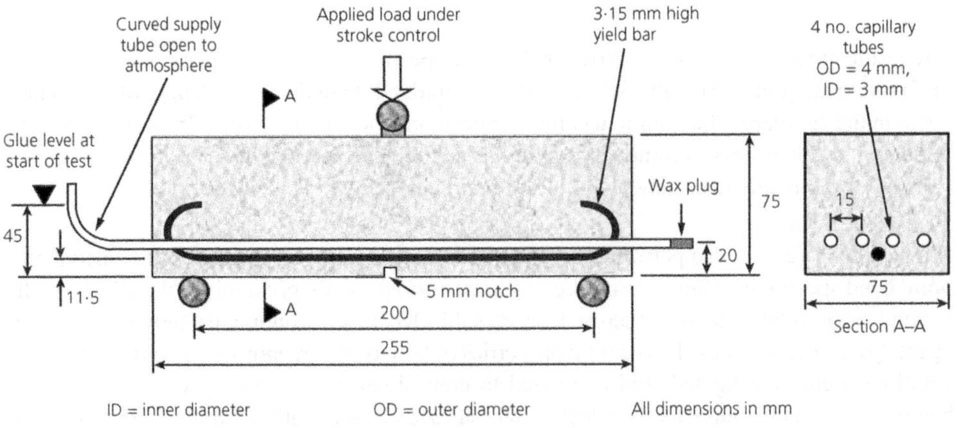

Figure 5 Experimental arrangement

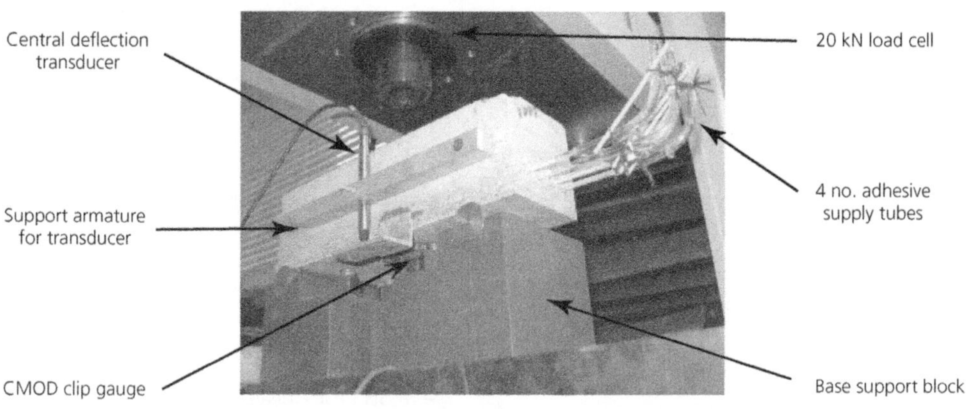

Figure 6(a) shows a typical load–CMOD response for a notched SH beam during the first and second (denoted SH beam (healed)) three-point bend tests, compared to that of a control beam. On breaking, the glass capillary tubes release cyanoacrylate into the crack plane which flows rapidly, under the influence of attractive capillary forces, across the two crack faces. The subsequent rapid curing of the cyanoacrylate in the moist alkaline environment of the mortar results in an almost instantaneous 'primary' healing effect.

The authors have also shown that the stiffness of the primary healing response increases with increasing reinforcement percentage and decreases with increasing loading rate (Joseph *et al.*, 2010). These responses are believed to be due to the rate at which the macrocrack opens. For higher levels of reinforcement, or slower loading rates, the rate of crack opening is slower and therefore the adhesive, which has flowed into the crack, and is in the process of hardening, is effectively being loaded at a slower rate. Joseph *et al.* (2010) also show qualitative evidence that supports the quantitative mechanical results obtained from their experiments. These include the sounds of glass tubes breaking, the presence of glue on the underside of the crack, glue-staining patterns on the final cracked faces, and clear evidence of new crack formation between the first and second loading cycles, as shown in Figure 6(b).

The autonomic healing system developed by Joseph *et al.* (2010) offers a successful mechanism for restoring, and, in certain circumstances, enhancing the mechanical properties of the composite. The infiltration of the cyanoacrylate into, and around, the macrocrack is likely to also reduce the permeability, and, therefore, improve the durability of the new composite material. The rapid flow and curing ability of the low viscosity cyanoacrylate, which is evident in the primary healing strength gain, also suggests that the method might be applicable to healing damage created under dynamic situations, such as in earthquakes.

Figure 6 (a) Typical Load–CMOD response for a notched SH and control beam; (b) new crack formation on side face of SH beam

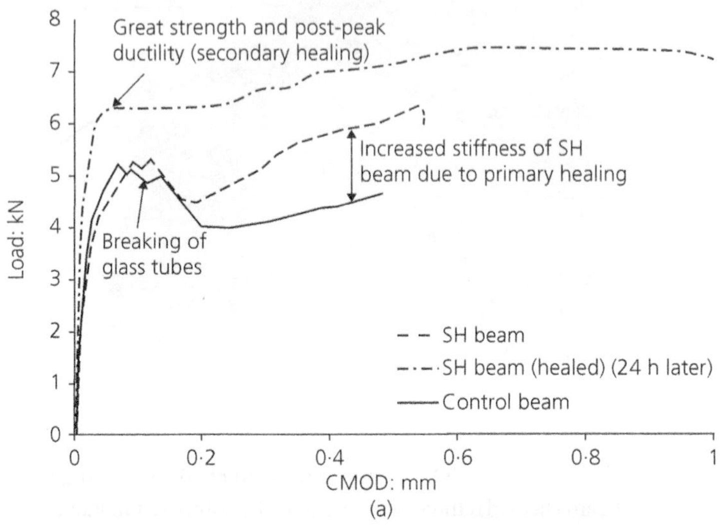

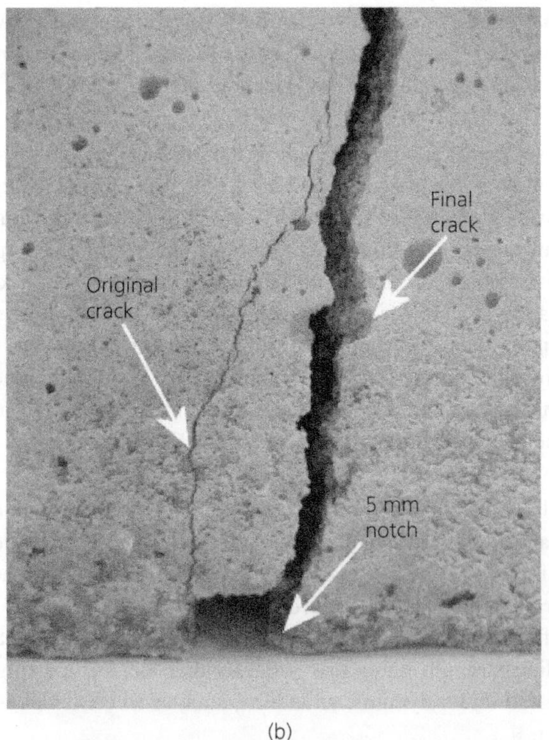

Further results demonstrating the feasibility of the autonomic healing of cementitious materials are presented by Dry (1994, 1996a, 1996b, 1996c, 1996d, 2000); Dry and Corsaw (2003); and Dry and Unzicker (1998). These studies consider various self-healing applications, including glass tube supply networks in loaded small-scale, rigid-framed structures, grids of scored glass tubes in large-scale concrete decks subject to shrinkage cracking, and hollow adhesive-filled fibres in reinforced concrete three-point bending samples. The majority of the results obtained from these studies are, however, largely qualitative, and are therefore not discussed here.

3.3 Autogenic healing in cementitious materials

3.3.1 Background

The autogenic healing of cementitious materials is a natural self-repair phenomenon. There are several processes that are considered to be responsible for this phenomenon. These include chemical, physical and mechanical interactions as discussed by Kishi *et al.* (2007). The main processes are: (*a*) swelling and hydration of cement pastes; (*b*) precipitation of calcium carbonate crystals; and (*c*) blockage of flow paths due to deposition of water impurities or movement of concrete fragments that detach during the cracking process.

The autogenous healing effect is generally acknowledged as one of the reasons why so many old buildings and structures have survived for so long with limited servicing and maintenance. Westerbeek (2005) attributes the unexpected longevity of many old bridges in Amsterdam to this phenomenon. It is believed that this longevity is due to the high levels of chalk or calcium in the cement of that area. Under the presence of water, this calcium is believed to dissolve and then deposit in cracks, thus partially healing them and hindering their propagation.

Over recent years many authors have continued to investigate this natural self-healing ability: Reinhardt and Joos (2003) have examined the effect of temperature on permeability and self-healing of cracked concrete; Zhong and Yao (2008) have investigated the effect of the degree of damage on the self-healing ability of normal-strength and high-strength concrete; Şahmaran and Li (2008) have considered the effect of autogenous healing on engineered cementitious materials (ECCs); and Jacobsen and Sellevold (1996) have examined the efficacy of autogenic healing on strength recovery of 'well cured' concrete beams exposed to rapid freeze–thaw cycles. The Jacobsen and Sellevold paper concluded that only a 4–5% recovery of compressive strength by means of autogenous healing was possible. Edvardsen (1999) noted, however, that the greatest potential for autogenous healing exists in early-age concrete.

More recent work by Ter Heide *et al.* (2005), as overviewed by Ghosh (2009), has therefore focused on examining both the mechanical strength gain and reduction in permeability of early-age concrete which has been cracked and allowed to heal autogenously. The authors examined the extent of healing under a range of factors including the specimen age at the time of cracking, cement type, width of the crack opening, compressive strength applied during healing, and the curing conditions post-damage. During the healing period, the specimens were subjected to various compressive stresses ranging from 0 to 2 N/mm^2 using spring-loading devices, in order to examine the effect of forced crack closure on the healing process.

The main conclusions from this study were that crack healing was only observed to occur for specimens stored under water, and that the amount of healing (strength recovery) was greatest for specimens which were initially damaged at an early age. The main process responsible for the healing behaviour was therefore believed to be continued ongoing hydration. In addition, the compressive stress applied to the crack faces was found to be very beneficial in respect to closing the initial crack, which was typically 50 μm wide. The authors note, however, that additional compressive forces above that required to cause crack closure and reinstate contact between the crack faces did little to improve the strength recovery following healing.

An interesting recent development has been the autogenous healing of expansive concretes as studied by Japanese researchers, namely Hosoda *et al.* (2007), Kishi *et al.* (2007) and Yamada *et al.* (2007). They have found, through microscopic observations and subsequent water permeability tests, that the inclusion of expansive agents in the concrete has allowed even large cracks of up to 0·3–0·4 mm to heal (Hosoda *et al.*, 2007). The authors have also found that the addition of small amounts of various carbonates such as bicarbonate of soda increase the self-healing ability of the concrete by allowing more calcium carbonate ($CaCO_3$) to be precipitated (Yamada *et al.*, 2007).

3.3.2   Enhancing autogenous healing behaviour

The use of compressive forces to close cracks and to create contact between crack faces has been shown to be an effective method of enhancing the natural autogenous healing process within cementitious materials, as discussed above (Ghosh, 2009; Ter Heide *et al.*, 2005). The method of crack closure through the external application of compressive forces is, however, impractical for most concrete members, particularly those within in situ structures. Recent work undertaken at Cardiff University has therefore examined the feasibility of low-level post-tensioning of cementitious materials using shrinkable polymer tendons (Isaacs *et al.*, 2010; Jefferson *et al.*, 2010).

The system involves the incorporation of unbonded pre-oriented polymer tendons in cementitious beams. Post-tensioning is achieved by thermally activating the shrinkage mechanism of the restrained polymer tendons after the cement-based material has undergone initial curing. The basic concept for the system is illustrated in Figure 7.

The concept has been investigated in the present work, using tendons formed from shrinkable polyethylene terephthalate (PET). These were employed in a series of three-point bending experimental tests on small-scale hollow prismatic mortar beams. These beams were pre-cracked and then subjected to heating in order to activate the shrinkage mechanism

Figure 7 Schematic illustration of concept for new composite material system

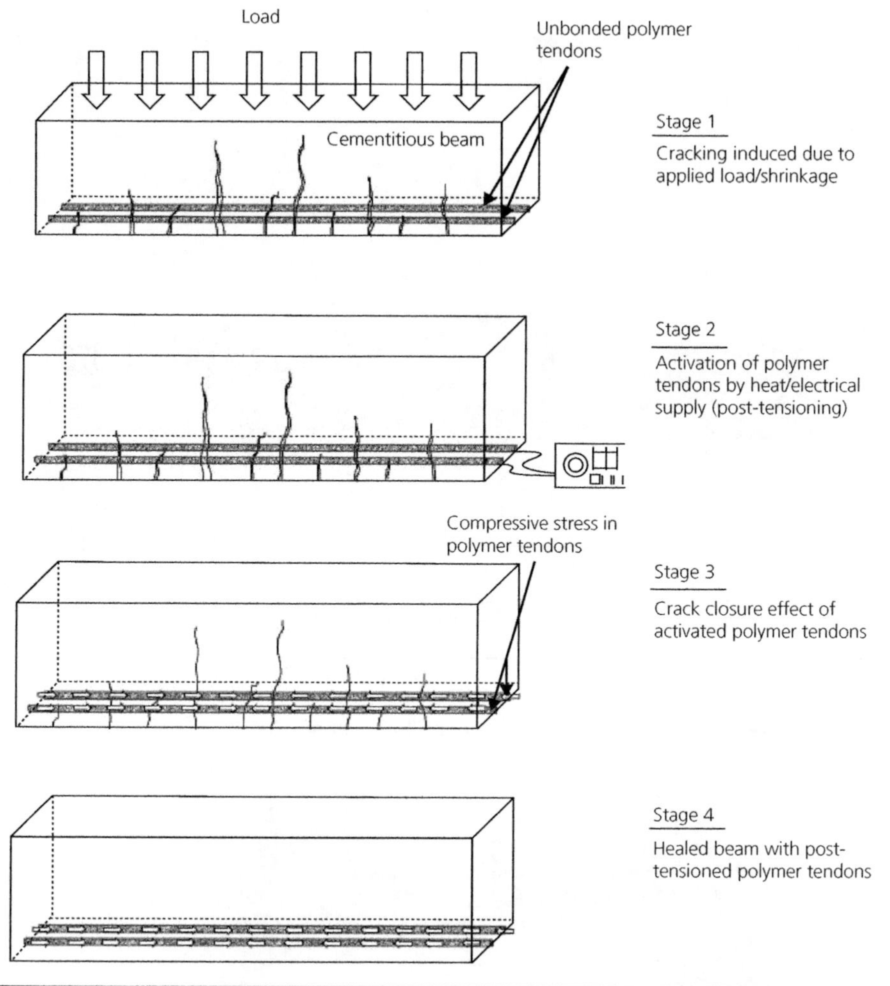

within the polymer tendons. The testing configuration used for the proof of concept tests is shown in Figure 8.

After preparation, the tests were conducted in three stages: stage 1 being loading to form a crack of 0·3 mm CMOD (day 4), stage 2 being heating at 90°C for 18 h to activate the PET shrinkage followed by a subsequent healing period (days 4–8), and stage 3 being reloading (day 8). The load–CMOD response was recorded for all specimens at stages 1 and 3 using the three-point bending testing arrangement illustrated in Figure 9. Prior to the second three-point bending test on day 8, half of the specimens containing the PET tendon had their tendon removed. Conducting the day 8 tests on specimens with and without the PET tendon remaining enables two parameters to be examined. Firstly, the magnitude of post-tensioning

Figure 8 Testing configuration for proof of concept tests for post-tensioning of cementitious materials using shrinkable polymer tendons

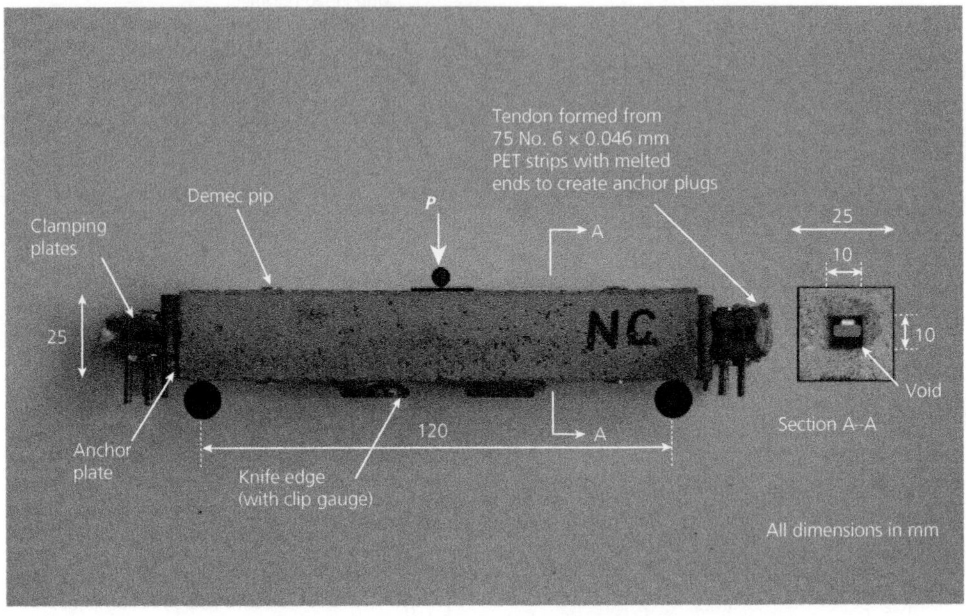

Figure 9 Three-point bending testing arrangement (Isaacs *et al.*, 2010)

effect provided by the shrinkage of the PET tendons on the mortar beams can be examined, and secondly, the benefit of subjecting the damaged beams to this post-tensioning for the duration of the healing period (day 4–8) can be examined explicitly.

Jefferson *et al.* (2010) have shown that the shrinkage activation of the PET tendon is sufficient to completely close the 0·3 mm crack created during the initial three-point loading on day 4, as shown in Figure 10. Furthermore, the reloading test data on day 8 indicate that not only has the

Figure 10 (a) A 0·3 mm wide crack after initial loading on day 4; (b) crack completely closed following shrinkage activation of PET tendon in oven

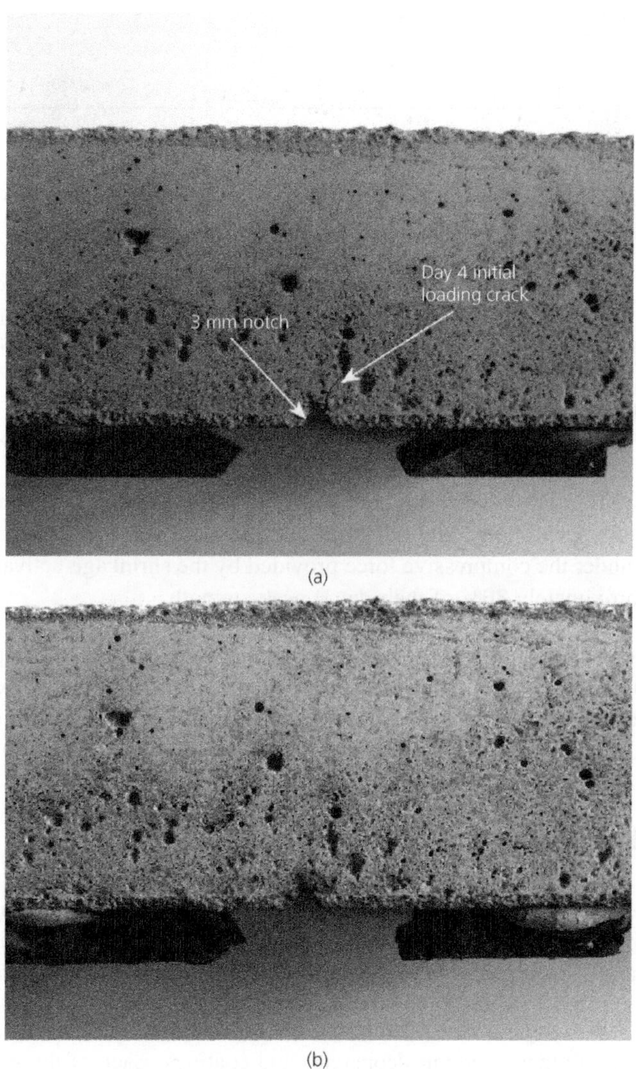

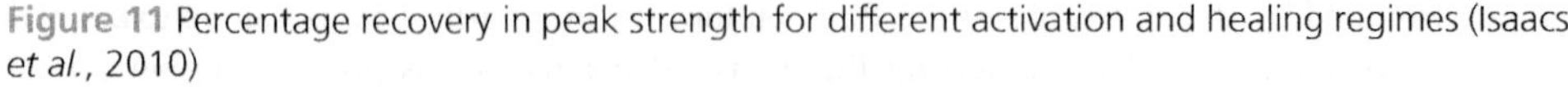

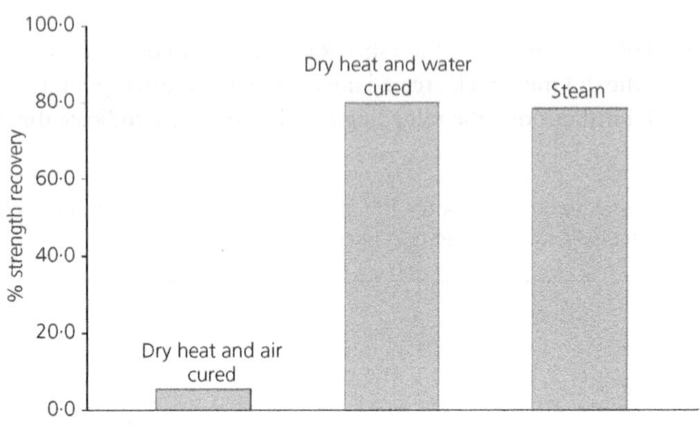

restrained shrinkage of the tendon caused the crack within the mortar beam to completely close but it has also put the beam into compression.

In more recent work (Isaacs *et al.*, 2010) the authors have examined the explicit effect that the post-tensioning provided by the PET tendon has on the degree of autogenous healing achieved between days 4 and 8. Various polymer activation and mortar healing conditions have been investigated including: (*a*) thermal activation in oven (90°C for 18 h) followed by healing in air for 3 days; (*b*) thermal activation in oven (90°C for 18 h) followed by healing in water at ambient temperature for 3 days; and (*c*) steam activation and healing in steam environment for 3 days.

It can be seen from Figure 11 that despite being loaded to a CMOD of 0·3 mm at day 4, and effectively being cracked completely through, the mortar specimens, when healed in either water or steam under the compressive force provided by the shrinkage-activated PET tendon, can recover approximately 80% of their day 4 peak strength.

The authors also highlight the fact that this new composite shape memory polymer cementitious system also has the potential to offer crack prevention in addition to crack healing, if the integral polymer tendons are activated prior to the occurrence of early-age cracking (Isaacs *et al.*, 2010). In addition, the possible use of recycled PET for the post-tensioning tendons also reduces the environmental footprint of this new composite material.

4. Conclusions

This paper has presented a review of current work in the arena of self-healing cementitious materials. Self-healing materials, which have the ability to respond and adapt to their environment, thereby improving their long-term durability and service life, potentially offer huge savings over their lifetime, both financially and in terms of their environmental impact.

There is a very wide range of research being undertaken at present on self-healing materials, including metals, polymers, ceramics/concretes and coatings. Each of these materials has its own potential niche application to the construction industry and each should be judged on its own merits for given applications. The development of self-healing concrete, however, potentially offers the greatest benefit to the construction industry over the coming years. Considering that over 6 billion m^3 of concrete is used annually, and that the majority of its durability problems originate from cracking, then the development of a smart cementitious material with the ability to self-heal cracks offers potentially massive savings to the annual amount of money spent on repair and maintenance of concrete structures.

The feasibility of the brittle glass delivery system is clearly the biggest drawback of the autonomic-healing method, in respect of on-site structural applications. Recent self-healing research within sandwich structures (Williams *et al.*, 2007) and polymers (Wu *et al.*, 2007) has moved towards the development of continuous supply networks that allow healing agent reserves to be replenished frequently. In light of this, two possible supply systems, which would improve the future outlook of the autonomic healing process in concrete, are suggested in Figure 12 (Joseph, 2008).

The natural autogenic healing of cementitious materials offers a more immediately feasible solution to crack repair. Autogenous healing has, however, only usually been able to offer a relatively low degree of strength recovery and has only been able to repair fine cracks in the order of 50 μm or less. Recent research has shown, however, that for early-age cracked mortar specimens, which are subject to compressive crack closure forces, an 80% recovery in the peak strength can be obtained (Isaacs *et al.*, 2010).

It seems that it is the development of novel composite materials, such as the post-tensioning of cementitious materials using shrinkable polymers, that are the most readily applicable self-healing materials to the construction industry. Not only do they have the ability to automatically

Figure 12 Future options for development of SH delivery system in concrete: (a) ductile network with ability to turn brittle post-casting; and (b) two-part supply tube which prevents blockages (Joseph, 2008)

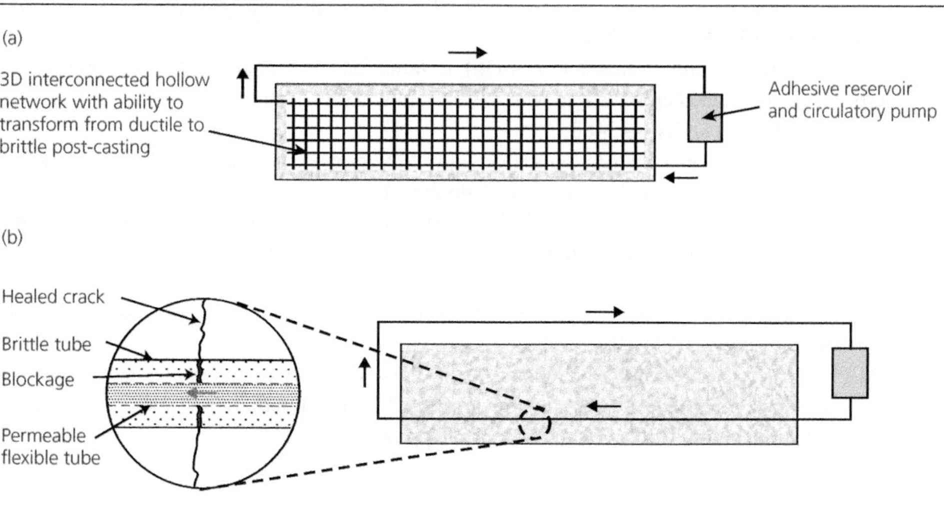

close cracks but also to significantly enhance the autogenous healing and thus long-term durability of cementitious materials. It is envisaged that such materials can be used in applications ranging from water-retaining structures to heavy civil structures that are exposed to the environment and hence the deleterious agents which promote the deterioration of concrete. Furthermore, due to the anticipated reduction in maintenance and repair of the self-healing materials, they immediately have benefit over traditional cementitious materials in situations where access to the structure is difficult such as airport runways and aprons, and where maintenance regimes are costly.

The development and use of any self-healing materials will inevitably involve some additional costs in the short term. However, providing the long-term efficacy of the system in respect to improving the materials, durability performance is proven, these initial costs should be more than offset by the long-term savings from reduced repair and maintenance. It is therefore in the interest of the construction industry to support the development of self-healing materials, particularly where the potential long-term savings to the industry, due to increased durability, are greatest; for example, self-healing concrete. Furthermore, the industry and clients must be prepared to evaluate the whole-life costs, both financially and environmentally, when specifying construction materials for civil engineering projects.

REFERENCES

Brown EN, Sottos NR and White SR (2002) Fracture testing of a self-healing polymer composite. *Experimental Mechanics* **42(4)**: 372–379.

Brown EN, Kessler MR, Sottos NR and White SR (2003a) In situ poly(urea-formaldehyde) microencapsulation of dicyclopentadiene. *Journal of Microencapsulation* **20(6)**: 719–730.

Brown EN, Moore JS, White SR and Sottos NR (2003b) Fracture and fatigue behaviour of a self-healing polymer composite. *Bioinspired Nanoscale Hybrid Systems*, Materials Research Society, Warrendale, PA, **735** pp. 101–106.

Brown EN, White SR and Sottos NR (2004) Microcapsule induced toughening in a self-healing polymer composite. *Journal of Materials Science* **39(5)**: 1703–1710.

Brown EN, White SR and Sottos NR (2005a) Retardation and repair of fatigue cracks in a microcapsule toughened epoxy composite – Part I: Manual infiltration. *Composites Science and Technology* **65(15–16)**: 2466–2473.

Brown EN, White SR and Sottos NR (2005b) Retardation and repair of fatigue cracks in a microcapsule toughened epoxy composite – Part II: In situ self-healing. *Composites Science and Technology* **65(15–16)**: 2474–2480.

Corporatewatch (2004) The construction sector: a brief overview. See http://www.corporatewatch.org/?lid=262 for further details (accessed 04/06/2009).

Dry CM (1994) Matrix cracking repair and filling using active and passive modes for smart timed release of chemicals from fibres into cement matrices. *Smart Materials and Structures* **3(2)**: 118–123.

Dry CM (1996a) Procedures developed for self-repair of polymeric matrix composite materials. *Composite Structures* **35(3)**: 263–269.

Dry CM (1996b) Release of smart chemicals for the in-service repair of bridges and roadways. *Proceedings of Smart Materials, Structures, and MEMS.* The International Society for Optical Engineering, Bangalore, India, December, 3321, pp. 140–144.

Dry CM (1996c) Smart bridge and building materials in which cyclic motion is controlled by internally released adhesives. *Proceedings of Smart Structures and Materials 1996: Smart Systems for Bridges, Structures, and Highways, Society of Photo-Optical Instrumentation Engineers, San Diego, CA, USA,* February, 2719, pp. 247–254.

Dry CM (1996d) Smart earthquake-resistant materials: using time-released adhesives for damping, stiffening, and deflection control. *Proceedings of the 3rd International Conference on Intelligent Materials and 3rd European Conference on Smart Structures and Materials, Society of Photo-Optical Instrumentation, Lyon, France,* June, 2779, pp. 958–967.

Dry CM (2000) Three designs for the internal release of sealants, adhesives, and waterproofing chemicals into concrete to reduce permeability. *Cement and Concrete Research* **30(12)**: 1969–1977.

Dry C and Corsaw M (2003) A comparison of bending strength between adhesive and steel reinforced concrete with steel only reinforced concrete. *Cement and Concrete Research* **33(11)**: 1723–1727.

Dry C and Unzicker J (1998) Preserving performance of concrete members under seismic loading conditions. *Proceedings of Smart Structures and Materials 1998: Smart Systems for Bridges, Structures, and Highways, The International Society for Optical Engineering, San Diego, CA, United States,* March, 3325, pp. 74–80.

Edvardsen C (1999) Water permeability and autogenous healing of cracks in concrete. *ACI Materials Journal* **96(4)**: 448–454.

Ghosh SK (ed.) (2009) *Self-healing Materials: Fundamentals, Design Strategies, and Applications.* Wiley-VCH, Weinheim, Germany.

Hosoda A, Kishi T, Arita H and Takakuwa Y (2007) Self-healing of crack and water permeability of expansive concrete. *Proceedings of the 1st International Conference on Self-healing Materials, Noordwijk, Holland,* April. Cd-rom.

Isaacs B, Jefferson AD, Joseph C, Lark RJ and Dunn S (2010) Autogenous healing in an SMP cementitious material system. Submitted to *Cement and Concrete Composites.*

Jacobsen S and Sellevold EJ (1996) Self-healing of high strength concrete after deterioration by freeze/thaw. *Cement and Concrete Research* **26(1)**: 55–62.

Jefferson AD, Joseph C, Lark RJ, Isaacs B, Dunn S and Weager B (2010) A new system for crack closure and low-level post-tensioning of cementitious materials using shrinkable polymers. *Cement and Concrete Research* **40(5)**: 795–801.

Joseph C (2008) *Experimental and Numerical Study of the Fracture and Self-healing of Cementitious Materials.* PhD thesis, Cardiff University.

Joseph C, Jefferson AD, Isaacs B and Lark RJ (2010) Experimental investigation of adhesive-based self-healing of cementitious materials. *Magazine of Concrete Research,* in press.

Kessler MR, Sottos NR and White SR (2003) Self-healing structural composite materials. Composites Part A. *Applied Science and Manufacturing* **34(8)**: 743–753.

Kishi T, Ahn T, Hosoda A, Suzuki S and Takaoka H (2007) Self-healing behaviour by cementitious recrystallisation of cracked concrete incorporating expansive agent. *Proceedings of the 1st International Conference on Self-healing Materials, Noordwijk, Holland,* April. Cd-rom.

Li VC, Yun Mook L and Yin-Wen C (1998) Feasibility study of a passive smart self-healing cementitious composite. *Composites Part B: Engineering* **29(6)**: 819–827.

Mihashi H, Kaneko Y, Nishiwaki T and Otsuka K (2000) Fundamental study on development of intelligent concrete characterized by self-healing capability for strength. *Transactions of the Japan Concrete Institute* **22(2000)**: 441–450.

Nissan Motor Co (2004) Nissan press release. See http://www.nissan-global.com/EN/NEWS/ 2005/_STORY/051202-01-e.html for further details (accessed 01/06/2009).

Office for National Statistics (2008) *Construction Statistics Annual Report 2008*. Palgrave Macmillan, Source: UK Statistics Authority website: http://www.statistics.gov.uk/ (accessed 05/06/2009).

Reinhardt H-W and Joos M (2003) Permeability and self-healing of cracked concrete as a function of temperature and crack width. *Cement and Concrete Research* **33(7)**: 981–985.

Şahmaran M and Li VC (2008) Durability of mechanically loaded engineered cementitious composites under highly alkaline environments. *Cement and Concrete Composites* **30(2)**: 72–81.

Schlangen E (2005) Self-healing phenomena in cement-based materials. *RILEM*. See http://www. rilem.net/tcDetails.php?tc=221-SHC for further details (accessed 01/06/2009).

Ter Heide N, Schlangen E and van Breugel K (2005) Experimental study of crack healing of early age cracks. *Proceedings of Knud Højgaard Conference on Advanced Cement-Based Materials,* Technical University of Denmark, June. Cd-rom.

Van der Zwaag S (ed.) (2007) *Self-healing Materials: an Alternative Approach to 20 Centuries of Material Science,* 1st edn. Springer, the Netherlands.

Westerbeek T (2005) Self-healing materials. Radio Netherlands. See http://www2.rnw.nl/rnw/en/ features/science/050801rf?view=Standard for further details (accessed 10/11/2005).

White SR, Sottos NR, Geubelle PH, *et al.* (2001) Autonomic healing of polymer composites. *Nature* **409(6822)**: 794–797.

Williams HR, Trask RS and Bond IP (2007) Design of vascular networks for self-healing sandwich structures. *Proceedings of the 1st International Conference on Self-healing Materials, Noordwijk, Holland,* April. Cd-rom.

Wu W, Hansen C, Aragon A, Sottos NR, White SR, Geubelle P and Lewis J (2007) Direct ink writing of microvascular networks. *Proceedings of the 1st International Conference on Self-healing Materials, Noordwijk, Holland,* April. Cd-rom.

Yamada K, Hosoda A, Kishi T and Nozawa S (2007) Crack self-healing properties of expansive concretes with various cements and admixtures. *Proceedings of the 1st International Conference on Self-healing Materials, Noordwijk, Holland,* April. Cd-rom.

Zhong W and Yao W (2008) Influence of damage degree on self-healing of concrete. *Construction and Building Materials* **22(6)**: 1137–1142.

Dhir and Paine
ISBN 978-0-7277-6457-7
https://doi.org/10.1680/icetsc.64577.049
ICE Publishing: All rights reserved

Chapter 4

Assessing the self-healing capability of cementitious composites under increasing sustained loading

Gurkan Yildirim
PhD Candidate, Department of Civil Engineering, Gazi University, Ankara, Turkey

Ahmed Alyousif
PhD Candidate, Department of Civil Engineering, Ryerson University, Toronto, ON, Canada

Mustafa Şahmaran
Associate Professor, Department of Civil Engineering, Gazi University, Ankara, Turkey

Mohamed Lachemi
Professor, Department of Civil Engineering, Ryerson University, Toronto, ON, Canada

This study investigated the effects of progressively increasing sustained loading on self-healing behaviour of 180-d-old microcracked engineered cementitious composites (ECCs) incorporating different mineral admixtures. After introducing microcracks to the specimens with applied severe pre-loading, some were subjected to progressively increasing sustained loading. All of the specimens were then subjected to continuous moist curing for 150 d to evaluate self-healing performance. Mechanical property (modulus of rupture (MOR) and mid-span beam deflection) characterisations and ultrasonic pulse velocity measurements were used to assess self-healing capability. Experimental results showed that even under progressively increasing sustained mechanical loading, MOR results greater than the original values could be obtained, depending on mineral admixture selection. Although deflection results were more adversely affected by progressive sustained loading compared to MOR results, even the lowest deflection value obtained from different ECCs was still more than 100 times that of conventional concrete after healing. Under continuous moist curing, there were minimal changes in ultrasonic pulse velocity results of all ECCs subjected to progressively increasing sustained loading, so that recovery results similar to those of specimens without sustained loading were obtained, despite the fact that ultrasonic pulse velocity testing was not that sensitive in capturing the effects of self-healing.

Introduction

Concrete is designed and used for its high compressive strength in most construction practices. Recently, however, other characteristics such as durability have started to come to the forefront. Durability of concrete material is generally accepted to be closely related to crack formation, which is most probably the result of material brittleness. Moreover, strength grade increments and packing of mixtures do not contribute to prevention of brittleness; adversely, they can exacerbate it, which significantly increases the chance of cracking, especially at early ages. Although awareness of material brittleness and its negative effects on durability have been raised by researchers, the design of concrete mixtures accounting for high brittleness has remained in the background because the tools needed to design concrete with controlled cracking and crack width are limited. However, recent efforts have made it possible to manufacture high-performance fibre-reinforced cementitious composites (HPFRCC) counteracting the potential drawbacks of high brittleness (Naaman and Reinhardt, 2003).

Ductal is a notable type of HPFRCC (Bache, 1981). It uses a design approach in which a tightly packed matrix for increased strength is used in combination with fibres for high ductility. Since the matrices in the mixtures are designed to be dense, strong bonding is established with the fibres themselves, which allows high post-cracking strength, given the usage of fibres with high strength. Under such circumstances, the material is able to reach 12 MPa tensile strength and 0·02–0·06% ductility (Chanvillard and Rigaud, 2003). As another emerging class of HPFRCC, engineered cementitious composites (ECCs) were first introduced over the last two decades (Li, 1998). Unlike Ductal, design of ECCs concentrates on the maximisation of tensile ductility through synergistic interactions among individual components (i.e. fibres, matrix and the interface between the two) and usage of the least possible fibre amount (2% or less, by volume). The superior tensile ductility of ECCs, which is a consequence of strain-hardening behaviour, starts to be visible as multiple, micron-sized, flat microcracks (<100 µm) propagate over the specimens, keeping crack widths constant as crack lengths increase to at least 100 times (3–5%) those of conventional concrete commonly used in the field (Li, 2003). Moreover, the formation of such narrow microcracks allows for the inclusion of special attributes (i.e. high self-healing capability) into the material (Bertagnoli *et al.*, 2011).

Inherent self-healing in concrete (the closing of cracks with newly formed products over time) is a widely studied phenomenon in the research community, with records and studies dating back to nineteenth century (Hearn and Morley, 1997; Hyde and Smith, 1889). Despite the knowledge surrounding the mechanism, studies have recently started to grow in number due to the increasing cost of repeated repair and/or maintenance of deteriorating infrastructures, as well as the invention of novel materials that make self-healing possible. According to Li and Herbert (2012), self-healing approaches for concrete-like materials fall into five categories: chemical encapsulation, bacterial encapsulation, mineral admixtures, chemicals in glass tubing and intrinsic self-healing, which is favoured by tight cracking. From a practical implementation point of view, the natural self-healing of tight cracks with no external interference is the simplest approach, despite the novelty of all of the above-mentioned approaches. Therefore, many studies focusing on the effects of intrinsically well-controlled tight crack widths in ECCs and their relation to natural self-healing mechanisms have been conducted. These studies have investigated the self-healing performance of ECCs under a number of commonly encountered environments, using pre-loaded specimens in the unloaded state. They concluded that durability, transport and mechanical properties can be recovered up to a great extent under certain conditions (Lepech and Li, 2005; Li *et al.*, 2007; Sahmaran *et al.*, 2014; Sahmaran and Li, 2008, 2009; Yang *et al.*, 2009). However, upon unloading, cracks were almost half the width of those on loaded specimens (Yang *et al.*, 2007; Zhou *et al.*, 2010), which led to the overestimation of self-healing performance. This point is also important considering the so-called creeping effect, which causes further opening of existing cracks with extended service lives. Therefore, the influence of sustained loading on self-healing performance is of value, especially when more conservative and realistic results are needed. In this sense, a comprehensive study was previously undertaken by the authors to assess the effect of constant sustained loading and different environmental exposures on mechanical property recovery of ECCs (Ozbay *et al.*, 2013). That paper concluded that even under constant sustained flexural loading, significant self-healing of microcracks took place. Although the sustained loading was monitored continuously, slight relaxations in stress values were inevitable under constant loading. Hence, to account for possible relaxation under constant stress values and to make the situation even more severe, the current study focuses on the effects of progressively increasing sustained loading and different mineral admixture utilisation on the self-healing capability of ECCs. Mechanical loading was applied after 180 d of initial curing, which was more than enough for almost complete maturity of cementitious systems packed with high levels of mineral admixture. Results were evaluated by ultrasonic pulse velocity (UPV) and mechanical property measurements of specimens.

Experimental programme

Materials, mixture proportions and basic mechanical properties

Four different ECC mixtures incorporating three different types of fly ash (two Class F and one Class C) and ground granulated blast furnace slag (S, (ECC_S)) were produced. Although two of the fly ash types were selected to be Class F in accordance with ASTM C 618 (ASTM, 2003), denomination was made according to the lime content in mixtures incorporating fly ash (FA) (i.e. low calcium (FA-L, (ECC_L)), medium calcium (FA-M, (ECC_M)) and high calcium (FA-H, (ECC_H))). In addition to four mineral admixtures (MAs), ordinary Portland cement (PC), silica sand with maximum aggregate size of 400 μm, water, polyvinyl alcohol (PVA) fibres with a diameter of 39 μm, nominal tensile strength of 1610 MPa, specific gravity of 1·3 and high-range water-reducing admixture (HRWRA) were used in the mixtures. Table 1 presents the chemical and physical properties of Portland cement and the different MAs. ECC mixtures were produced with a

Table 1 Chemical and physical properties of PC and different MAs

Chemical composition: %	PC	FA-L	FA-M	FA-H	S
Silicon dioxide (SiO_2)	20·8	57·0	55·0	41·8	35·1
Aluminium oxide (Al_2O_3)	5·6	21·0	22·3	18·8	10·4
Iron (III) oxide (Fe_2O_3)	3·4	4·2	4·0	6·4	0·79
Magnesium oxide (MgO)	2·5	1·8	1·2	4·7	12·2
Calcium oxide (CaO)	61·4	9·8	11·9	21·8	38·3
Sodium oxide (Na_2O)	0·19	2·2	2·4	1·7	0·12
Potassium oxide (K_2O)	0·77	1·5	0·80	0·63	0·37
Loss on ignition	2·2	1·3	1·8	0·68	–
Silicon dioxide + aluminium oxide + iron (III) oxide	29·8	82·2	81·3	67·0	46·3
Physical properties					
Specific gravity	3·06	2·02	2·59	2·61	2·87
Blaine fineness: m^2/kg	325	290	306	315	430

water-to-cementitious materials (PC+MA) ratio (w/cm) of 0·27, and mineral admixture to Portland cement ratio (MA/PC) of 1·2, by mass. Mixture proportions are listed in Table 2. Basic mechanical properties of ECC mixtures were assessed considering the flexural parameters under four-point bending load (flexural strength and mid-span beam deflection) and the compressive strength results of 28, 90 and 180-d-old $360 \times 75 \times 50$ mm prism (length $\times$ width $\times$ depth) and 50 mm cubic specimens, respectively. After the completion of casting for mechanical property characterisation, specimens were kept in moulds for 1 d at $23 \pm 2°C$, $50 \pm 5\%$ relative humidity (RH) and then in plastic bags at $23 \pm 2°C$, $95 \pm 5\%$ RH for 6 d, followed by curing in a laboratory medium at $23 \pm 2°C$, $50 \pm 5\%$ RH until the end of 28, 90 and 180 d. Mechanical property characterisations were made using the average of results obtained from eight different specimens and the values are included at the end of Table 2. As seen from Table 2, while the 28-d-old compressive strength results of fly ash-bearing ECCs were in the range of 55·0–59·0 MPa; ECC_S specimens reached 77·2 MPa, due to higher fineness and cementing behaviour of individual slag particles. Variation in flexural strength results of ECC mixtures was not as high as for the compressive strength results, varying between 9·9 and 11·6 MPa after 28 d. This behaviour is related to more elaborate material properties (e.g. tensile first cracking strength, ultimate tensile strength and tensile strain capacity) governing flexural strength results (Qian *et al.*, 2009). ECC_S mixture showed the lowest average mid-span beam deflection after 28 d (2·4 mm), while ECC_L mixture was almost two times that of the ECC_S specimens with 4·4 mm. Increased ductility of ECC_L specimens was attributed to the tendency of

Table 2 Mixture proportions and basic mechanical properties of ECC mixtures

Mixture proportions		Mix ID			
		ECC_L	ECC_M	ECC_H	ECC_S
Portland cement		1·0	1·0	1·0	1·0
Sand		0·80	0·80	0·80	0·80
FA/PC		1·2	1·2	1·2	–
S/PC		–	–	–	1·2
w/cm		0·27	0·27	0·27	0·27
PVA: kg/m^3		26	26	26	26
HRWRA: kg/m^3		3·9	4·2	4·6	5·3
Mechanical properties					
Compressive strength: MPa	28 d	55·5	54·7	58·7	77·2
	90 d	66·2	69·5	61·9	81·5
	180 d	72·6	73·4	69·8	84·3
Flexural strength: MPa	28 d	10·2	9·9	11·6	11·5
	90 d	10·7	10·6	11·7	11·9
	180 d	10·8	10·9	11·6	12·1
Flexural deflection: mm	28 d	4·4	3·8	3·1	2·4
	90 d	4·2	3·2	3·0	2·4
	180 d	4·1	3·1	2·9	2·2

low-calcium fly ash particles to reduce the PVA fibre/matrix interface chemical bond and matrix toughness, while increasing the interface frictional bond in favour of attaining high tensile strain capacity (Wang and Li, 2007). When 90- and 180-d-old basic mechanical properties of ECCs are considered, it is evident that, despite changes in the values compared to 28-d-old specimens, results did not show marked changes between individual curing ages, implying that 90 d of curing was almost adequate for full maturity attainment.

Sample preparation and pre-cracking

As previously stated, the main purpose of this study was to investigate the effect of progressively increasing sustained loading on self-healing of microcracked ECCs. To do this, a number of prism specimens ($360 \times 75 \times 50$ mm (length $\times$ width $\times$ depth)) were produced from each mixture. After 24 h inside the moulds at $23 \pm 2°C$, $50 \pm 5\%$ RH, all of the specimens were removed and put inside plastic bags at $23 \pm 2°C$, $95 \pm 5\%$ RH for 180 d. A 180-d curing period was selected to make sure of the completion of most of the hydration reactions and high maturity. As discussed previously, attainment of very high maturity after 180 d of initial curing can clearly be monitored from mechanical property results. For observation of self-healing in damaged ECCs, microcracks were produced in the specimens using four-point bending loading in a controlled manner at the age of 180 d. Since it is not easy to create similar microcracking damage on ECC mixtures incorporating MAs with greatly varying chemical compositions, the initial pre-loading level of specimens was selected to be 75% of the mid-span beam deflection values at 180 d. 180-d-old average mid-span beam deflection results of at least six different specimens from ECC_L, ECC_M, ECC_H and ECC_S mixtures were 4·1, 3·1, 2·9 and 2·2 mm, respectively, as seen in Table 2. After the application of initial pre-loading, some of the pre-loaded beam specimens underwent progressively increasing sustained loading for an additional 150 d, while others did not. For the specimens kept in the sustained loading device, after initial pre-loading, the level of the first sustained loading was set to 40% of ultimate flexural load. After each 30-d period under water, the sustained loading level was increased by 10% until 120 d (80%) was reached (Figure 1). A torque wrench was used to control periodic increments in sustained flexural loading (Figure 2). A pair of steel rollers supported the beams and loading was applied with similar rollers placed over one-third

Figure 1 Changes in sustained loading level with time

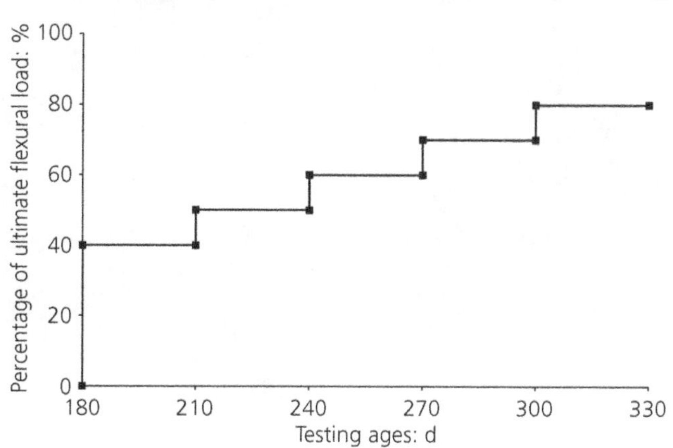

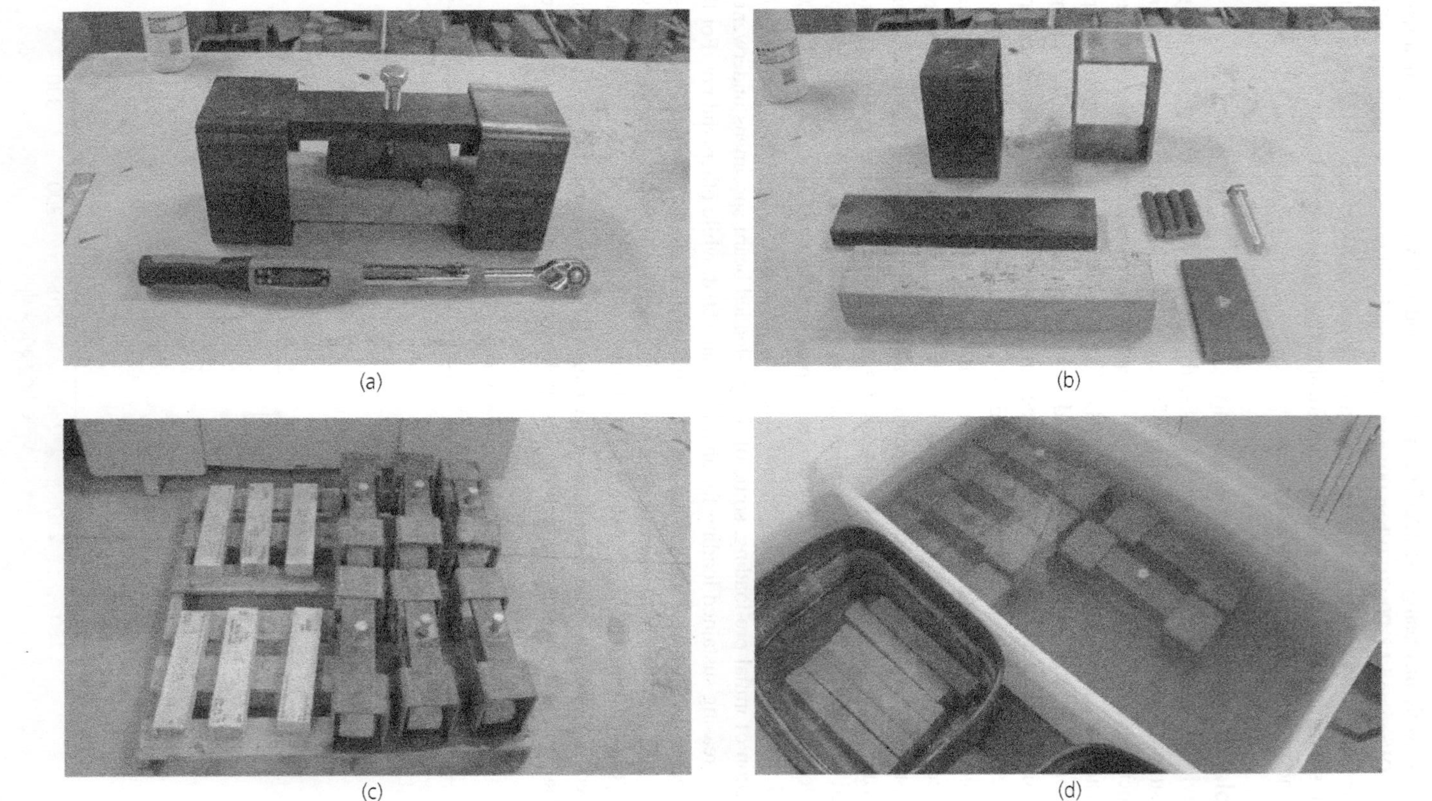

Figure 2 Torque wrench, application of sustained loading and curing of specimens

Figure 3 Calibration of torque wrench against the load cell 2 weeks before the start of actual tests

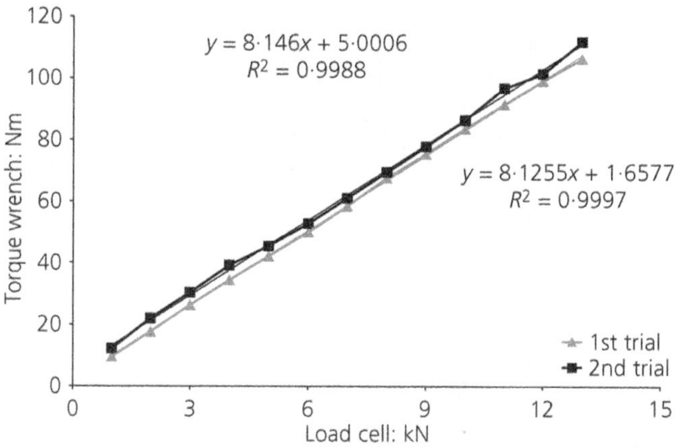

of the supporting span length. Specimens were loaded using threaded rods tightened by a torque wrench. To evaluate the load placed on the specimens, a load cell was initially inserted in place of the specimen and torque was applied with the torque wrench, which was then calibrated against the load cell. This calibration was determined by way of two separate trials before testing (Figure 3). As seen from Figure 3, there is a linear relationship between the applied torque and loading itself, and results correspond well with each other. It must be mentioned that during each 30-d period of sustained loading, slight relaxations took place, although these were accounted for in the curing periods of each sustained loading level. Along with the pre-loaded specimens, some beam specimens without pre-loading (sound) were kept for comparison. To evaluate self-healing, longitudinal ultrasonic pulse velocity (UPV) measurements were performed on specimens with and without progressively increasing sustained loading after each 30-d period of water exposure, according to ASTM C 597 (ASTM, 2009). Moreover, mechanical properties (modulus of rupture (MOR) and mid-span beam deflection) of ECCs with and without sustained loading were investigated before and after 150 d of moist curing.

Results and discussion

Recovery of mechanical properties

Modulus of rupture

Modulus of rupture results, calculated by using the maximum flexural loads from load–deflection curves, are displayed in Table 3. All the values presented in this table were calculated by averaging at least eight different beams tested under four-point bending loading. The table includes the results of sound ECC specimens tested up to failure after 180 d of initial curing and after 150 d of further water curing beyond the first 180 d (330 d), together with specimens pre-loaded up to 75% of their mid-span beam deflection capacities on their 180th day, then released and again re-loaded up to failure. Moreover, the effect of progressively increasing sustained loading on self-healing capability was monitored, with specimens pre-loaded up to 75% of their mid-span beam deflection capacities after 180 d of initial curing and cured for 150 d under water, with or without progressively increasing sustained loading. Considering 180-d MOR results of sound ECCs with

Table 3 Flexural properties of ECC mixtures before and after continuous water exposure

Mix ID	Sound specimens tested up to failure at 180 d		Sound specimens tested up to failure at 330 d		180-d-old sound specimens pre-loaded, released and re-loaded until failure		Specimens pre-loaded after 180 d and exposed to 150 d of continuous moist curing			
							Without sustained loading		With sustained loading	
	MOR: MPa	Deflection: mm	MOR: MPa	Deflection: mm	MOR: MPa	Deflection: mm	MOR: MPa	Deflection: mm	MOR: MPa	Deflection: mm
ECC_L	11·1	4·1	11·4	3·7	10·4	2·6	11·2	3·8	11·0	3·5
ECC_M	11·2	3·1	11·6	3·0	9·9	2·0	10·8	2·7	10·6	2·4
ECC_H	11·8	2·9	12·0	2·7	9·3	1·8	11·1	2·3	10·3	1·9
ECC_S	12·1	2·2	12·2	2·1	9·0	1·5	11·2	2·0	10·0	1·6

different MAs, it can clearly be stated that results were close to each other, ranging between 11·1 and 12·1 MPa levels. Despite the minor differences in results, the highest average MOR values were acquired from the ECC_S specimens, while the lowest were from ECC_L. This finding regarding the results of mixtures incorporating different MAs is believed to be in strict connection with the previously discussed factors affecting compressive strength.

When focus is placed on the effect of pre-loading on MOR results, it is clear that pre-loading up to 75% of the mid-span beam deflection capacities of specimens resulted in decrements that differed from the initial values by 6·3%, 11·6%, 23·7% and 25·6% for ECC_L, ECC_M, ECC_H and ECC_S mixtures, respectively. These findings show that the effectiveness of pre-loading in lowering MOR results varied and that the ductility of different mixtures was a decisive parameter in damage tolerance. To put it differently, initial pre-loading was less harmful in terms of MOR results for specimens with higher ductility. This was an anticipated outcome since some amount of loss in load-carrying capacity is inevitable upon pre-loading, regardless of MA type. In addition, cracks re-open from the same place with load application, despite significant crack closure upon load removal. However, one important point requiring further attention is that, after pre-loading, minimum MOR results were achieved by ECC_S (9·0 MPa) specimens while the maximum results were obtained in ECC_L (10·4 MPa), which differs from the results of sound specimens tested up to failure with no initial pre-loading. This result suggests that slag-incorporated ECCs are more sensitive to pre-loading compared to those with fly ash (especially class F fly ash), resulting in less recovery upon removal of the existing load. As suggested above, this can be related to the reduced sensitivity of mixtures with lower deflection capacity (i.e. ductility) against loading. However, it is important to note that although all of these specimens almost failed with initial pre-loading up to 75% deflection capacity, they were able to recover most of their flexural strength values in the unloaded state, with the minimum average value reaching 9·0 MPa at the age of 180 d.

Although the effect was rather restricted, there were increments in the MOR results of 180-d-old sound specimens as they were aged for an additional 150 d under water. As indicated in Table 3, increases in MOR results were less pronounced in slag-bearing ECCs compared to those produced with fly ash (especially class F fly ash). This behaviour was due to the fact that 180 d of initial curing was more than enough for almost complete stabilisation of hydration reactions of highly reactive ECC_S specimens and the additional 150 d of moist curing did not make a noticeable difference. On the other hand, fly ash particles were more likely to contribute to improved MOR results, even after 330 d of complete curing, owing to the highly pozzolanic behaviour of fly ash compared to slag, which manifested itself more in later ages.

Considering the effect of self-healing on specimens subjected to pre-loading up to 75% of their mid-span beam deflection values after 180 d and kept without sustained loading for a further 150 d under water, it can be seen that recoveries resulting in values close to or even slightly higher than that of sound specimens are achievable through proper selection of MA type. For example, while MOR results of sound ECC_L, ECC_M, ECC_H and ECC_S specimens were 11·1, 11·2, 11·8 and 12·1 MPa after 180 d of curing, initial pre-loading and subsequent 150 d of water exposure resulted in final values of 11·2, 10·8, 11·1 and 11·2 MPa, respectively. This indicates the efficacy of self-healing on MOR results with no regard to MA type. These results show that after 150 d under water, ECC_L specimens attained average results greater than the sound specimens. Moreover, ECC_M specimens exhibited the second-highest recovery ratio compared to sound specimens. In the case of the ECC_S mixture, however, pre-loaded specimens showed the lowest recovery level in comparison with sound specimens. These results suggest that final self-healing products of ECCs with class F fly ash were stronger than ECC_S specimens, considering the recovery of MOR results. Additionally, these findings offer an insight into the nature of final self-healing products, which were also visually monitored on the crack faces of different specimens. Figure 4 shows that after a certain amount of continuous

Figure 4 View of microcracked surfaces after exposure to continuous moist curing

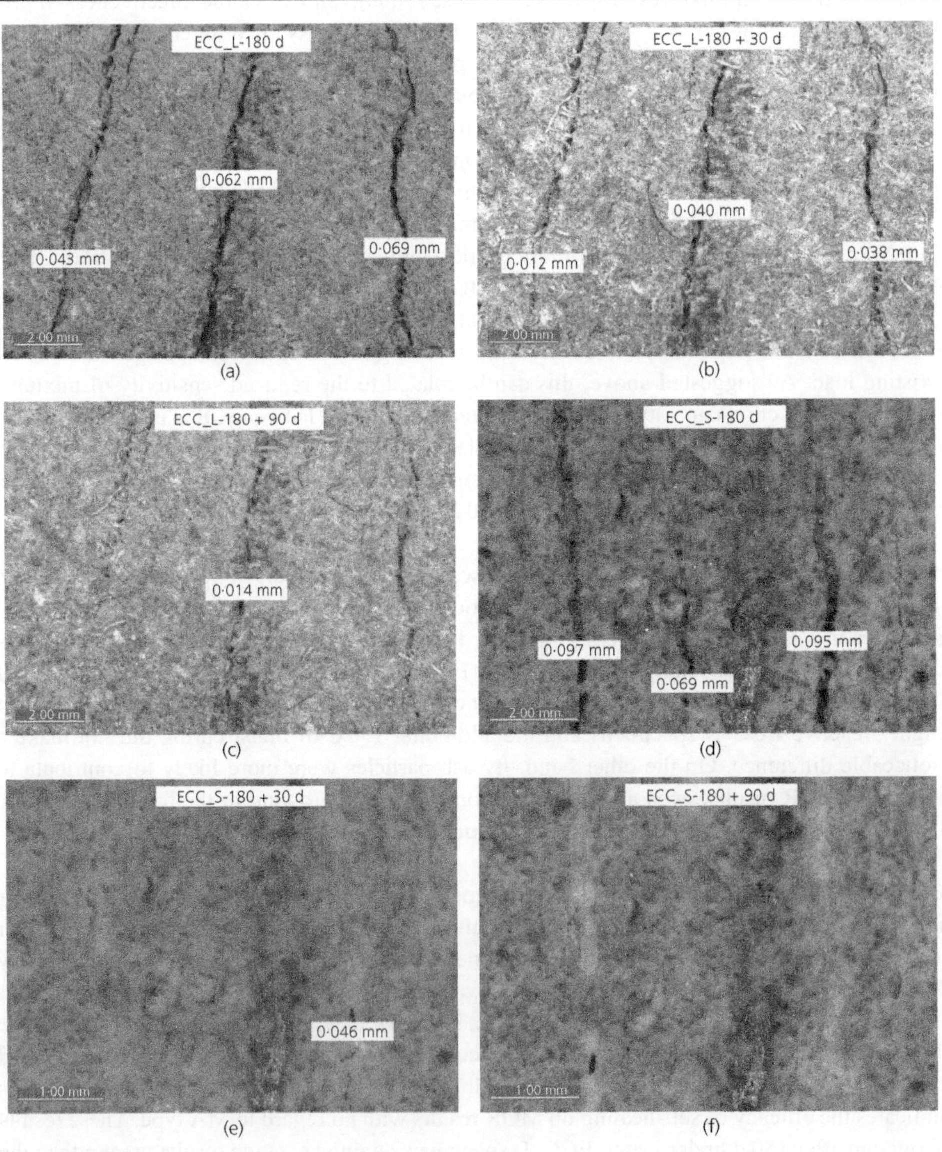

(a)   (b)   (c)   (d)   (e)   (f)

water exposure of pre-loaded specimens, cracks of ECC_S specimens were covered with traces of calcite, manifesting itself as white residue. This was not the case in specimens with fly ash, which suggests that further-formed calcium–silicate–hydrate (C–S–H) gels were the dominant self-healing products, and these are considerably stronger than calcite (Sahmaran *et al.*, 2013). Overall, these findings could be the reasons why higher MOR recovery rates were obtained from ECCs with fly ash than those with slag.

The effect of progressively increasing sustained loading on the recovery of MOR results was another point requiring further inspection. As seen in Table 3, the application of sustained loading, with intensity increased by 10% of ultimate flexural load after each 30-d period until the end of 120 d, slightly lowered the MOR results in comparison to pre-loaded specimens that were not subjected to sustained loading. For instance, 330-d MOR results of ECC_H mixtures were 11·1 MPa when no sustained loading was applied. However, under progressive sustained loading, the result decreased to 10·3 MPa. This trend was valid for all mixtures produced in this study, and it was anticipated to be attributable to re-opening of microcracks after increments of sustained loading levels at 30-d intervals. However, it is notable that with careful selection of MA, almost complete recovery of MOR results (according to sound specimens) is possible, even after severe initial pre-loading and subsequent exacerbation. It is important that, despite having slightly lower recovery rates than specimens without sustained loading, the values obtained from specimens with sustained loading are still close to original values (180-d sound specimens), and higher than the specimens that failed directly after initial pre-loading. As for the specimens without sustained loading, ECC_S specimens showed the lowest recovery in MOR results compared to ECCs with fly ash (especially class F FA (ECC_L and ECC_M)) when they were subjected to progressive sustained loading. Probable explanations for ECC_S behaviour could be: very limited pozzolanic capacity of slag particles reducing the chance for self-healing as time passes; the tendency for slag-bearing specimens to exhibit cracks with larger widths due to enhanced matrix toughness and chemical bonding between PVA fibres and matrices (Sahmaran *et al.*, 2012); and the lower capability of newly formed calcite particles to withstand monthly increased sustained loading compared to C–S–H gels.

Mid-span beam deflection

Table 3 presents the mid-span beam deflection results of different ECC mixtures. Flexural deflection results reflecting the materials' ductility were calculated by using at least eight different beam specimens from each mixture. During the calculation of deflection results, values corresponding to maximum flexural load points on load–deflection curves were taken into consideration, and residual flexural deflection from pre-loading was not included in the overall results for more reasonable estimation. Table 3 shows substantial differences in deflection results based on different MA usage, with 180-d results of sound specimens showing a variation between 2·2 and 4·1 mm levels. When sound specimens were further cured under water for 150 d, slight reductions in deflection results were monitored due to continuous improvement of matrix and fibre/matrix interface properties; the same results ranged between 2·1 and 3·7 mm levels after this time. Considering the results of sound specimens from both ages, ECC_S specimens displayed the lowest mid-span beam deflection levels, with ECC_L specimens showing the highest. Potential causes of this finding have been discussed in previous sections.

When initial pre-loading was applied to the specimens before sudden re-loading, there were significant decrements in the final deflection results. For example, while average deflection capacity of 180-d-old sound ECC_L specimens was 4·1 mm, the value went down to 2·6 mm with applied pre-loading; this situation held true for the rest of the mixtures. Moreover, decrement rates in deflection results were higher compared to MOR values. For instance, the rate of decrease in MOR values of ECC_L specimens was 6·3% after pre-loading, although the rate increased to 36·6% in mid-span beam deflection results. This might be attributed to faster localisation of already-introduced individual microcracks, which cause final failure without significantly sacrificing maximum flexural load upon re-loading.

Exposure to 150 d of moist curing after initial pre-loading without any sustained loading led to marked improvements in mid-span beam deflection results, no matter what type of MA was used. When a comparison is made between specimens subjected to pre-loading at the age of 180 d, then released and re-loaded up to failure, and those subjected to pre-loading and left to self-heal for 150 d under water without any sustained loading, recoveries up to 46·1% were observed with appropriate MA usage (from 2·6 to 3·8 mm in ECC_L specimens). Similar MOR results were observed in specimens produced with slag and class C fly ash, with lower recovery rates compared to those incorporating class F fly ash. Reasons for the differing behaviours of different mixtures have been discussed in previous sections. In addition to results of pre-loaded specimens, results obtained after conditioning were close to those of sound specimens directly tested up to failure after 330 d of initial curing. For example, average deflection results of 330-d-old sound specimens after failure were 3·7, 3·0, 2·7 and 2·1 mm for ECC_L, ECC_M, ECC_H and ECC_S mixtures, and 3·8, 2·7, 2·3 and 2·0 mm for the same mixtures kept in water for 150 d after severe pre-loading. Considering the results of specimens subjected to progressively increasing sustained loading after initial pre-loading on the 180th day and moist curing for 150 d, deflection results showed minor reductions compared to specimens without sustained loading; values dropped to 3·5, 2·4, 1·9 and 1·6 mm for ECC_L, ECC_M, ECC_H and ECC_S mixtures, respectively. Overall, these findings were also in line with the behaviour observed in MOR results. Despite slight decrements with applied progressive sustained loading, even the lowest deflection values (1·6 mm in ECC_S specimens) acquired under progressive sustained loading were still more than 100 times that of conventional concrete, which has an ultimate mid-span beam deflection value nearly 0·01 mm.

Recovery of ultrasonic pulse velocity measurements

ECCs without sustained loading

The UPV results for different ECC mixtures are provided in Table 4, which shows the values for specimens subjected to initial pre-loading up to 75% of their 180-d-old mid-span beam deflection capacities, and for specimens subjected to monthly increased sustained loading starting from the 180th day until the end of the 330th day after the initial pre-loading. Values were found by taking average readings of eight beam specimens from each mixture. Table 4 clearly shows the differences in UPV test results of sound ECCs with different MAs. Results for 180-d UPV were calculated by averaging results of sound specimens tested with and without progressive sustained loading; the results were 3954, 4018, 4005 and 4253 m/s for ECC_L, ECC_M, ECC_H and ECC_S mixtures, respectively. These results indicate a close relationship with enhanced matrix maturity and high UPV results.

Table 4 Ultrasonic pulse velocity (UPV) test results of ECC specimens (units in m/s)

Mix ID		180 d			180+30 d		180+60 d		180+90 d		180+120 d		180+150 d	
				40%[c]		50%		60%		70%		80%		–
		Sound	PL[a]	PL+SL[b]	PL	PL+SL	PL	PL+SL	PL	PL+SL	PL	PL+SL	PL	PL+SL
ECC_L	W/o SL	3965	3888	–	4036	–	4058	–	4080	–	4114	–	4121	–
	With SL	3943	3831	3763	–	4014	–	4022	–	4067	–	4063	–	4101
ECC_M	W/o SL	4004	3822	–	4098	–	4109	–	4131	–	4168	–	4192	–
	With SL	4031	3922	3874	–	4001	–	4028	–	4093	–	4126	–	4179
ECC_H	W/o SL	4002	3841	–	4113	–	4194	–	4213	–	4256	–	4268	–
	With SL	4008	3866	3835	–	4069	–	4140	–	4079	–	4206	–	4283
ECC_S	W/o SL	4264	3999	–	4357	–	4365	–	4352	–	4402	–	4423	–
	With SL	4241	3996	3944	–	4286	–	4281	–	4332	–	4357	–	4430

[a] Pre-loaded.
[b] Pre-loaded and sustained loading.
[c] Numbers in percentages are the levels of sustained loading.

With the application of initial pre-loading, decrements in UPV results occurred regardless of MA type. For example, as specimens were initially pre-loaded, UPV results decreased by an average of 2·4, 3·6, 3·8 and 6·0% in sound specimens for ECC_L, ECC_M, ECC_H and ECC_S mixtures, respectively. These results show that, with the application of pre-loading, the highest decrements were obtained from ECC_S specimens in terms of UPV results as well. This finding has been previously discussed and attributed to the beneficial effects of slag on improved matrix maturity, which, in turn, increased the chemical bonding between fibres and matrix and reduced the formation of cracks with narrower widths. Although all of the specimens that were subjected to initial pre-loading were almost failed under four-point bending loading, percent changes in UPV results with the applied initial pre-loading were minimal and do not seem to show big variations, especially in the case of specimens with fly ash. This could be due to the similarities in chemical compositions of different MAs, as well as the fact that UPV tests are not sensitive enough to detect the changes in specimens after the introduction of microcracks. Moreover, it seems that UPV results are closely related to average crack width measurements, so in specimens with wider crack widths (e.g. ECC_S), the drop in UPV results is greater. It is important to note that, when considering average crack widths measured in different ECCs of the same age, ECC_S specimens showed the largest crack formation with 110–70 μm levels, whereas the same values were near 80–30 μm for ECCs with different types of fly ash after 180 d of initial curing.

In addition to the raw data in Table 4, Figure 5(a) shows the percentage variations in UPV measurements of ECC mixtures not subjected to progressively increasing sustained loading after severe initial pre-loading. This figure also reflects the monthly self-healing performance of beam specimens in terms of UPV measurements. According to Figure 5(a) and Table 4, it can easily be concluded that there were improvements in UPV measurements with time, irrespective of MA used in ECC mixtures showing self-healing up to a certain extent. Taking the ECC_S specimens as an example, average UPV results increased from 3999 to 4357 and 4423 m/s after 30 and 150 d of moist curing, respectively. In addition to the final UPV results after 330 d, values continuously increased after each 30-d interval and showed the formation of self-healing in microcracks. However, it should be noted that self-healing of microcracks upon exposure to continuous moist curing and continuing hydration reactions are not the only factors responsible for the increments in UPV results with time. Although the tests were started after 180 d of initial curing to reduce further hydration effects, the results were directly influenced by the integrity of the materials. Despite the varying changes in UPV results with further moist curing, 30 d under water was more than adequate for all ECC mixtures to attain UPV results greater than those of sound specimens. For example, UPV results of sound ECC specimens were 3965, 4004, 4002 and 4264 m/s after 180 d of initial curing for ECC_L, ECC_M, ECC_H and ECC_S specimens, respectively, while after severe initial pre-loading and 30 d of continuous moist curing, they were 4036, 4098, 4113 and 4357 m/s, respectively. In addition, most of the improvements in UPV results were achieved in the first 30 d of moist curing, although there was continuity until the end of the full 150-d period. This was most probably due to the fact that the anhydrous cementitious materials diminished significantly after 210 d of curing, lowering the chance for both crack healing and marked hydration.

An increasing trend in UPV results was valid for all ECC mixtures. Figure 5(a) shows a maximum 11·1% improvement in UPV results of initially pre-loaded specimens depending on

Figure 5 Percentage variations in UPV results of ECCs with the applied initial pre-loading and further curing: (a) without sustained loading; (b) with sustained loading

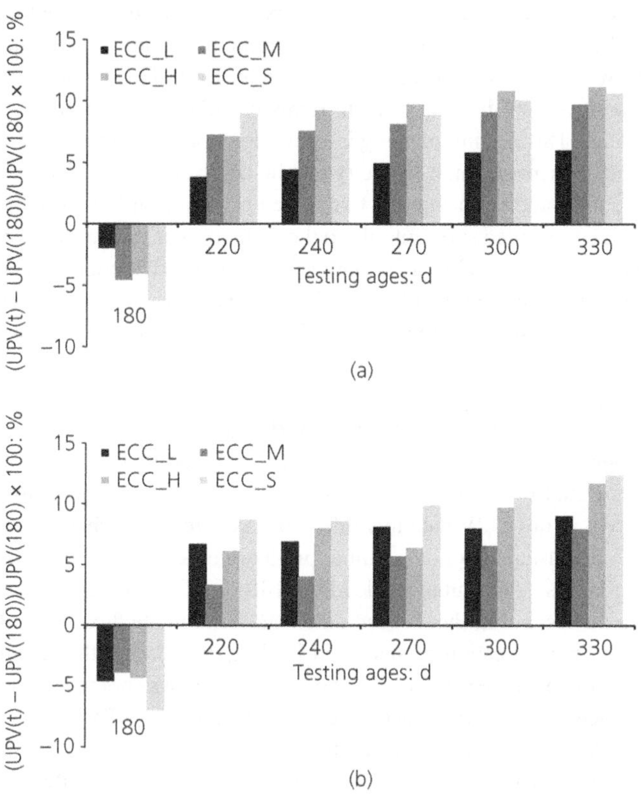

curing age and MA used in mixtures. Based on this figure, the higher UPV recovery results obtained from ECC_H and ECC_S mixtures in general had values that were comparable to each other at the end of each 30-d interval until the end of 150 d. ECC_M and ECC_L mixtures followed in terms of improvements in UPV results with further curing. This result does not correspond with the recovery of mechanical property results, and was unexpected since the superior pozzolanic capacity of class F fly ash particles is likely to contribute to the formation of self-healing and more mature paste, especially at later ages. In a different study by the authors of this paper, similar results showed the dominance of self-healing in ECCs with class C fly ash and, in particular, slag (Sahmaran *et al.*, 2013). That study reported that lower silicon dioxide (SiO_2) amounts present – especially in the composition of slag particles – could lower the chance for pozzolanic reactions, thereby increasing the amount of portlandite and, relatedly, the pH value of pore solution. When pore solution has higher pH values, carbonic acid (a product of reaction between water and carbon dioxide) can disassociate faster as bicarbonate (HCO_3^-) and carbonate ions (CO_3^{2-}), which are used in the formation of calcite by combining with calcium (Ca^{2+}) ions leached away from C–S–H gels and portlandite in the presence of carbon dioxide-abundant water (Edvardsen, 1999; Jooss, 2001). Thus, increased self-healing capability of ECC_S specimens was attributed to the formation of calcite up to a considerable extent, whereas specimens bearing class C fly ash were influenced by both calcite and C–S–H gel formation (Sahmaran *et al.*, 2013). The discrepancy in the self-healing behaviours of ECCs with different MAs is believed to be due to the varying natures of different tests. The precise detection of self-healing in terms of UPV measurements is not an easy task, owing most probably to the lower sensitivity of the test technique itself. This could be another possible reason behind the lower self-healing rates in terms of UPV results from ECC mixtures containing class F fly ash. 180 d of initial aging is quite a long period for most pozzolanic reactions to take place, and thus pozzolanic capacity might not have been adequate to show itself effectively at the final stage of curing due to reduced precision of UPV testing. However, it is highly likely that it contributed substantially to the formation of the further C–S–H gels, which densified the matrices. As a concluding remark related to the UPV improvement results of pre-loaded specimens, the final UPV results of different mixtures were found to be very close to each other, although all of the specimens showed self-healing performances with varying rates (Table 4).

ECCs with progressively increasing sustained loading

The effects of progressively increasing sustained flexural loading on self-healing capability of ECCs initially subjected to severe pre-loading up to 75% of their mid-span beam deflection capacities are shown as raw UPV data in Table 4 and as percentage changes in Figure 5(b). As in specimens with no sustained loading, initial pre-loading led to decreased UPV results in all 180-d-old ECC mixtures. After initial pre-loading, 180-d-old beam specimens were put into the sustained loading device and further loaded until 40% of ultimate flexural loads were reached. As seen from Table 4, application of sustained loading, together with the initial pre-loading, caused UPV results to go down even more, although the effect was not as intense as in initial pre-loading. For instance, while a 5·8% decrease took place in the UPV results after initial pre-loading of 180-d-old sound ECC_S specimens, the rate increased up to 7·0% with additional sustained loading. This behaviour generally applied to mixtures with different fly ash types as well, which was expected due to further opening of already-introduced microcracks. When Figure 5(b) is evaluated in detail, it can be concluded that comparable self-healing performances were acquired from specimens that were only subjected to initial pre-loading before moist curing, and those subjected to progressively increasing sustained loading after initial pre-loading. Moreover, the increasing trend in healing rates was mostly continuous, despite the fact that, after each 30-d interval of moist curing, the torque wrench was adjusted to 10% more than the initial 40%, until reaching 80% of ultimate flexural load after 300 d. Taking the ECC_S mixture as an example, the healing rates after 30, 60, 90, 120 and 150 d of moist curing were 8·7, 8·5, 9·8, 10·5 and 12·3%, respectively. These rates were obtained when the sustained loading levels were 40, 50, 60, 70 and 80% of the ultimate flexural loads of ECC_S specimens. Enhanced self-healing performance in terms of UPV measurements was surprising in the presence of continuously increased sustained loading, since it was presumed that application of higher rates of sustained loading in every 30-d interval would overshadow the beneficial effects of self-healing. This might be explained by the initial crack opening factor and its possible effects on self-healing performance. When cracks with smaller openings were considered, self-healing was anticipated to occur at a slower rate due to space restriction for new self-healing products and slower diffusion of necessary species (i.e. water and carbon dioxede) into smaller cracks.

When the sustained loading level was increased, however, cracks were likely to open more, leading to easier ingress of required substances for self-healing to accelerate and providing more space for final healing products. Although some mixtures showed enhanced improvements in UPV recovery results, it can generally be stated that there were reductions in the healing rates of certain mixtures (ECC_M and ECC_H) when sustained loading was applied. However, it should not be overlooked that all of the mixtures with different MAs under the effect of progressively increasing sustained loading reached UPV results greater than those of sound specimens after only 30 d of water exposure. Throughout 150 d of complete curing, results did not show significant variations from each other. Overall this finding also provided an insight into how sensitive UPV tests were to changes after self-healing.

Conclusions

This paper discusses the effects of progressively increasing sustained loading on self-healing behaviour of initially pre-loaded mature ECC specimens produced with different MAs having a wide variety of chemical compositions. During the experimental programme, self-healing was evaluated with UPV measurements and mechanical property (modulus of rupture (MOR) and mid-span beam deflection) characterisations. The following conclusions have been drawn based on the results.

- As a consequence of 150 d of continuous moist curing led, MOR results of initially pre-loaded, 180-d-old ECC specimens not subjected to progressive sustained loading were close to and/or higher than those of sound specimens, depending on the type of MA used in the mixtures. This finding was informative regarding the strength of final self-healing products that stem from different ECCs utilising varying MAs.
- Application of sustained loading, intensified monthly beyond severe initial pre-loading, did not cause significant changes in MOR results when specimens were subjected to moist curing for 150 d. Results similar to those of sound specimens were achievable, although recovery rates were slightly lower than specimens with no sustained loading.
- Deflection results were very close for specimens failed after 180 d of initial curing and those initially pre-loaded on the 180th day and exposed to 150 d of moist curing without any sustained loading. When a similar comparison was made with specimens subjected to progressive sustained loading, there were reductions in the recovery results. However, even the lowest deflection values acquired under progressive sustained loading (1·6 mm in ECC_S) were still more than 100 times those of conventional concrete.
- Initial pre-loading and subsequent exacerbation with sustained loading after 180 d did not make a significant difference to the original UPV results. This is believed to be due to lower precision of UPV testing in capturing the related data.
- Self-healing performance results in terms of UPV measurements were close to each other for specimens subjected only to initial pre-loading before moist curing, and for those subjected to progressively increasing sustained loading after initial pre-loading. Although UPV recovery rates were not that marked, 30 d of continuous moist curing was more than enough for all ECCs to achieve UPV results higher than those of the sound specimens.

Acknowledgements

The authors gratefully acknowledge the financial assistance of the Scientific and Technical Research Council (Tubitak) of Turkey provided under project MAG-112M876 and the Turkish Academy of Sciences, Young Scientist Award programme.

REFERENCES

ASTM (2003) ASTM C 618: Standard specification for coal fly ash and raw or calcined natural pozzolan for use mineral admixture in Portland cement concrete. ASTM International, West Conshohocken, PA, USA.

ASTM (2009) ASTM C 597: Standard test method for pulse velocity through concrete. ASTM International, West Conshohocken, PA, USA.

Bache H (1981) *Densified Cement/Ultra-fine Particle-based Materials*. Aalborg Portland, Aalborg, Denmark. CBL Report No. 40.

Bertagnoli G, Mancini G and Tondolo F (2011) Early age cracking of massive concrete piers. *Magazine of Concrete Research* **63(10)**: 723–736, http://dx.doi.org/10.1680/macr.2011.63.10.723.

Chanvillard G and Rigaud S (2003) Complete characterization of tensile properties of ductal UHPFRC according to the French recommendations. In *Proceedings of High Performance Fiber Reinforced Cement Composites (HPFRCC-4)*. (Naaman AE and Reinhardt HW (eds)). Rilem SARL, Bagneux, France, pp. 21–34.

Edvardsen C (1999) Water permeability and autogenous healing of cracks in concrete. *ACI Materials Journal* **96(4)**: 448–455.

Hearn N and Morley CT (1997) Self-healing property of concrete–experimental evidence. *Materials and Structures* **30(7)**: 404–411.

Hyde GW and Smith WJ (1889) Results of experiments made to determine the permeability of cements and cement mortars. *Journal of the Franklin Institute* **128(3)**: 199–207.

Jooss M (2001) Leaching of concrete under thermal influence. *Otto-Graf-Journal* **12**: 51–68.

Lepech M and Li VC (2005) Water permeability of cracked cementitious composites. *Proceedings of the 11th International Conference on Fracture, Turin, Italy*, pp. 20–25.

Li M, Sahmaran M and Li VC (2007) Effect of cracking and healing on durability of engineered cementitious composites under marine environment. *Proceedings of High Performance Fiber Reinforced Cement Composites (HPFRCC-5), Stuttgart, Germany*, pp. 313–322.

Li VC (1998) ECC – Tailored composites through micromechanical modeling. *Proceedings of Fiber Reinforced Concrete: Present and the Future Conference*. CSCE Press, Montreal, Canada, pp. 64–97.

Li VC (2003) On engineered cementitious composites (ECC): A review of the material and its applications. *Advanced Concrete Technology* **1(3)**: 215–230.

Li VC and Herbert E (2012) Robust self-healing concrete for sustainable infrastructure. *Advanced Concrete Technology* **10(6)**: 207–218.

Naaman AE and Reinhardt HW (2003) Setting the stage: Toward performance-based classification of FRC composites. In *Proceedings of High Performance Fiber Reinforced Cement Composites (HPFRCC-4)*. (Naaman AE and Reinhardt HW (eds)). Rilem SARL, Bagneux, France, pp. 1–4.

Ozbay E, Sahmaran M, Yucel HE *et al.* (2013) Effect of sustained flexural loading on self-healing of engineered cementitious composites. *Advanced Concrete Technology* **11(5)**: 167–179.

Qian S, Zhou J, De Rooij MR *et al.* (2009) Self-healing behavior of strain hardening cementitious composites incorporating local waste materials. *Cement and Concrete Composites* **31(9)**: 613–621.

Sahmaran M and Li VC (2008) Durability of mechanically loaded engineered cementitious composites under highly alkaline environments. *Cement and Concrete Composites* **30(2)**: 72–81.

Sahmaran M and Li VC (2009) Influence of microcracking on water absorption and sorptivity of ECC. *Materials and Structures* **42(5)**: 593–603.

Sahmaran M, Yucel HE, Demirhan S *et al.* (2012) Combined effect of aggregate and mineral admixtures on tensile ductility of engineered cementitious composites. *ACI Materials Journal* **109(6)**: 627–638.

Sahmaran M, Yildirim G and Erdem TK (2013) Self-healing capability of cementitious composites incorporating different supplementary cementitious materials. *Cement and Concrete Composites* **35(1)**: 89–101.

Sahmaran M, Yildirim G, Ozbay E *et al.* (2014) Self-healing ability of cementitious composites: Effect of addition of pre-soaked expanded perlite. *Magazine of Concrete Research* **66(7)**: 409–419, http://dx.doi.org/10.1680/macr.13.00250.

Wang S and Li VC (2007) Engineered cementitious composites with high-volume fly ash. *ACI Materials Journal* **104(3)**: 233–241.

Yang EH, Yingzi Y and Li VC (2007) Use of high volumes of fly ash to improve ECC mechanical properties and material greenness. *ACI Materials Journal* **104(6)**: 620–628.

Yang Y, Lepech MD, Yang EH *et al.* (2009) Autogenous healing of engineered cementitious composites under wet-dry cycles. *Cement and Concrete Research* **39(5)**: 382–390.

Zhou J, Qian S, Beltran MGS *et al.* (2010) Development of engineered cementitious composites with limestone powder and blast furnace slag. *Materials and Structures* **43(6)**: 803–814.

Dhir and Paine
ISBN 978-0-7277-6457-7
https://doi.org/10.1680/icetsc.64577.069
ICE Publishing: All rights reserved

Chapter 5

Self-healing performance of engineered cementitious composites under natural environmental exposure

Benny Suryanto
Assistant Professor, School of Energy, Geoscience, Infrastructure and Society, Institute of Infrastructure and Environments, Heriot-Watt University, Edinburgh, UK

Sam Alan Wilson
MEng student, School of Energy, Geoscience, Infrastructure and Society, Institute of Infrastructure and Environments, Heriot-Watt University, Edinburgh, UK

William John McCarter
Professor, School of Energy, Geoscience, Infrastructure and Society, Institute of Infrastructure and Environments, Heriot-Watt University, Edinburgh, UK

Thomas Malcolm Chrisp
Professor, School of Energy, Geoscience, Infrastructure and Society, Institute of Infrastructure and Environments, Heriot-Watt University, Edinburgh, UK

The self-healing performance of an engineered cementitious composite (ECC) exposed to the natural environment is presented. Fifteen dog-bone shaped ECC samples were preloaded after 14 d of curing and then placed outside in an open area at Heriot-Watt University (Edinburgh campus). Ultrasonic pulse velocity measurements were used to determine the rate and extent of the self-healing capabilities of the ECC in the natural environment. The results showed that, while the more highly damaged samples displayed the greatest decrease in ultrasonic velocity, they also displayed initial accelerated healing, which implies an increased quantity of individual cracks in the more damaged samples rather than an increase in individual crack widths. It was also found that the self-healing of microcracks in the ECC was robust. Narrow hairline cracks ($< 10\ \mu m$ width) healed in less than 6 d, while 20–30 μm wide cracks either partially or fully healed after 6 d of intermittent rainfall. Wider microcracks (40–75 μm) partially healed after 3 weeks of outdoor exposure.

Notation

d	distance between two transducers in ultrasonic pulse velocity (UPV) measurements (mm)
F_{28}	28 d compressive strength (50 mm cube)
F_{90}	90 d compressive strength (50 mm cube)
f_{t14}	14 d tensile strength
R_c	ratio of UPV recorded in control sample before and after being placed outdoors
R_n	normalised ratio of UPV
R_p	ratio of UPV recorded in a preloaded sample before and after being placed outdoors
t	measured transit time (µs)
v_c	UPV of control samples before being placed outdoors (m/s)
$v_{c,e}$	UPV of control samples during natural outdoor exposure (m/s)
v_p	UPV of preloaded samples before being placed outdoors (m/s)
$v_{p,e}$	UPV of preloaded samples during natural exposure (m/s)
ε_{t14}	14 d tensile strain capacity

Introduction

The deterioration of reinforced concrete infrastructure is a worldwide problem. In most developed countries, the situation is exacerbated by the fact that a significant percentage of infrastructure is now approaching, or has even passed, its service life. In the UK, for example, since 2000, expenditure on repair and retrofitting existing infrastructure has exceeded that for new developments (fib, 2008; McCarter *et al.*, 2010, 2012). This figure is expected to rise as more construction takes place and existing infrastructure continues to age. It is the brittle nature of concrete that is often found to be one of the major causes of deterioration, in particular cracking within the cover-zone of concrete structures (Vaysburd *et al.*, 2005).

Over the lifetime of a concrete structure, cracking is seen as being inevitable. Many factors can contribute to the development and propagation of cracks, including uneven settlement, restrained shrinkage, unforeseen weather conditions, poor design and unforeseen loading scenarios. These factors can interact to exacerbate the deterioration process. Consider, for example, bridge deck concrete under cyclic loading. Repeated application of loading can cause existing cracks in the deck (due to restrained shrinkage or excessive loading) to propagate and increase in width (Maekawa *et al.*, 2013). If not repaired, these cracks can allow the penetration of moisture, oxygen and chlorides to the reinforcing bars, initiating corrosion and leading to further cracking and spalling of the concrete cover (Long *et al.*, 2001). The deterioration process is therefore accelerated as a result.

It is paradoxical that while moisture is the root cause of many deterioration processes in reinforced concrete, it is an essential ingredient for cement hydration processes. However, another important yet less well-known role of moisture is the opportunity to allow regain of loss in performance of concrete due to cracking through a process called autogeneous healing (Dhir *et al.*, 1973; Neville, 2012). The ingress of moisture into a crack can bring a number of advantages (Hearn, 1998; Edvardsen, 1999) – it can promote the hydration of unhydrated or partially reacted cement particles within the crack, which can result in the recovery of mechanical properties (Li *et al.*, 2013), it can allow existing hydration products to swell and gradually seal the crack, and it can introduce impurities that help to seal a crack. However, in order for these processes to effectively occur, crack widths must be small (Reinhardt and Jooss, 2002; Reinhardt *et al.*, 2013; Snoeck *et al.*, 2014).

The work presented here directs attention to the use of an engineered cementitious composite (ECC) as a potential self-healing material. An ECC is a ductile concrete that has a controllable crack width (less than 0·1 mm under loading) and a tensile strain capacity in excess of 3%, which is more than two orders of magnitude greater than that of ordinary concrete (Li, 2003; Suryanto *et al.*, 2010). It is the non-brittle nature of ECC and its inherent narrow crack widths that make it possible to study its self-healing performance in the natural environment. Such material therefore has the potential to extend the life of ageing infrastructure and protect new developments with more durable materials, leading to more sustainable construction in the future.

Previous studies have demonstrated the robust performance of ECC in healing microcracks under controlled laboratory conditions (Kan *et al.*, 2010; Li and Yang, 2007; Qian *et al.*, 2009, 2010; Yang *et al.*, 2009, 2011). These works highlighted the facts that the self-healing process in ECC takes place more effectively in crack widths of less than 50 μm and that strength and stiffness recoveries are achievable. Sisomphon *et al.* (2013) found that the extent of healing was influenced by the concentration of carbonate ions in the absorbed water as samples subjected to cyclic wetting and drying with regularly refreshed water showed a faster healing effect.

Herbert and Li (2012) were the first to assess the performance of ECC in the natural environment. They demonstrated that precracked ECC samples were able to gradually heal and recover to 90% of their initial resonant frequency values after 1 month of outdoor exposure. In subsequent work they also found that a faster recovery was achievable using ECC samples with a smaller tuned average crack width (approximately 24–30 μm), with some showing a 90–96% recovery in resonant frequency after 1 week of exposure to the natural environment (Herbert and Li, 2013). Variation in weather conditions was cited as the primary reason for the difference. Temperature, humidity and rainfall intensity vary on a daily basis and it is the climatic conditions local to the material that are of importance in terms of self-healing capability. If ECC is to develop into a viable global construction material, research must be carried out to determine its performance when subjected to a range of climatic conditions.

Scotland, having high precipitation, offers a suitable environment to exploit the self-healing potential of ECC. In eastern Scotland, where this research was conducted, rainfall is distributed relatively evenly throughout the year, with autumn and early winter typically being the wettest periods. The mean monthly rainfall typically lies in the range 40–65 mm (Met Office, 2014). Intermittent rain and showers are typical, providing conditions that are favourable for self-healing. This research aimed to assess the self-healing performance of a newly developed ECC mixture (Pourfalah and Suryanto, 2013) exposed to the natural environment of eastern Scotland (Edinburgh) over a winter period.

Experimental programme

In total, 20 dog-bone shaped samples were cast. As summarised in Table 1, five were used to determine the mechanical properties under uniaxial tensile loading, five were precracked to 60% of the average strain capacity, five others to 30% of this value and the remaining five

Table 1 Test samples

Samples	Remarks
S8–S10, S19, S20	Tested to failure to obtain stress–strain curves
S1–S3, S11, S12	Precracked at 30% ultimate tensile strain
S4, S6, S13, S14, S18	Precracked at 60% ultimate tensile strain
S5, S7, S15–S17	Control samples for outdoor testing

samples served as control samples to monitor the influence of continued hydration. These 15 samples were then positioned outdoors in an open yard and thus exposed to the natural environment for a period of 3 months in winter. Five 50 mm cubes were also prepared to monitor compressive strength development – two cubes were tested at 28 d and the remaining three were tested at 90 d.

Test samples and instrumentation

Dog-bone samples of dimensions recommended by JSCE (2008) were prepared (Figure 1(a)). Tensile testing was performed using a 100 kN Instron 4206 machine (Figure 1(b)). Each sample was held at both ends using pneumatic grips and loading was then performed under a crosshead speed of 0·5 mm/min. Figure 2 presents the tensile stress–strain responses obtained from the dog-bone samples. Stresses were determined from load cell readings, while strains within the centre region of the samples were obtained from the average of two linear variable

Figure 1 Tensile tests: (a) schematic diagram of a dog-bone sample (dimensions in mm); (b) sample during uniaxial tensile testing

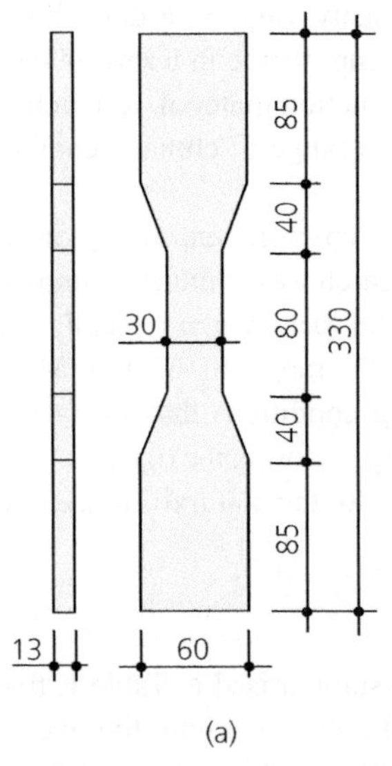

Figure 2 Typical tensile stress–strain response obtained from dog-bone shaped ECC samples

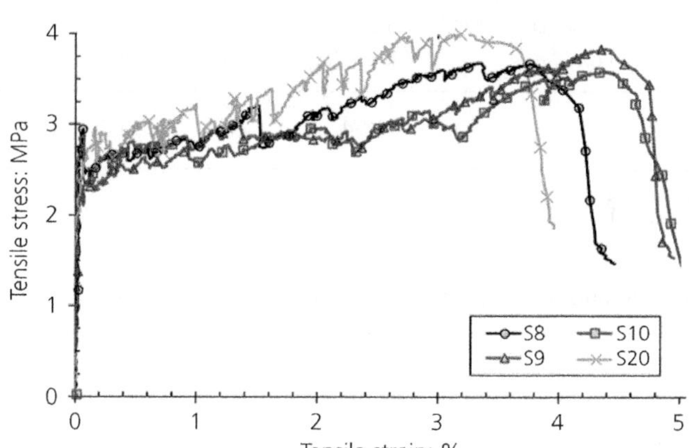

displacement transducer (LVDT) readings (Figure 1(b)). For the preloaded samples, loading was stopped when the tensile strain reached the desired value and the grips were released to unload the sample.

Ultrasonic pulse velocity (UPV) measurements were used to quantify the extent of initial damage and self-healing. A Pundit apparatus was used, together with two 54 kHz transducers. Intimate contact with the specimen surface was obtained using a viscous grease; in addition, prior to taking any measurements, calibration was carried out using a 26·0 μs reference bar. Measurements were undertaken by coupling the transducers to both ends of each sample (direct transmission measurement) and the signal transit time was recorded. These measurements were taken before and immediately after preloading, followed by twice a week over the first 2 weeks of outdoor exposure and then once a week throughout the remaining test period.

In addition to UPV measurements, a portable digital microscope equipped with a 12 MP digital camera (Veho VMS) was used to study the progress of self-healing in selected microcracks. Weather information was collected from a local weather station (Gogarbank), which is at a similar altitude and approximately 2 km away from the exposure site. Weather data were collected on a weekly basis from the WeatherCast website (WeatherCast, 2014).

Materials

Table 2 details the mix proportions used. The binder comprised CEM I 52·5N cement (Procem, Lafarge) and a fine fly ash (Superpozz SV80, Scotash) with a fly ash to cement (FA/C) ratio of 1·8. The water/binder (w/b) ratio was fixed at 0·28, which was found adequate for the given FA/C ratio, in order to provide a satisfactory matrix toughness and fibre–matrix interfacial strength to achieve strain hardening and controlled crack width, the latter being desirable for self-healing (Yang *et al.*, 2011). Fine silica sand with an average particle size of 120 μm (RH110, Minerals Marketing) was used in all mixes at a constant sand to cement ratio of 0·6 (by mass). Oxide analysis of the FA and silica sand is given in Table 3. A polycarboxylate high-range water-reducing (HRWR) admixture (Glenium C315, BASF) was added to the mix

Table 2 Mix proportions and mechanical properties

CEM I: kg/m^3	FA: kg/m^3	Silica sand: kg/m^3	w/b^a	HRWR: kg/m^3	PVA fibre: kg/m^3	F_{28}: MPa	F_{90}: MPa	f_{t14}: MPa	ε_{t14}: %
454	818	273	0·28	4·54	26	35·8	45·3	3·8	3·9

^{a} Binder includes cement and FA.

Table 3 Oxide analysis and physical properties of FA and silica sand (wt%)

	FA	Silica sand
Chemical analysis		
Silicon dioxide (SiO$_2$)	52·7	98·8
Aluminium oxide (Al$_2$O$_3$)	26·6	0·21
Iron oxide (Fe$_2$O$_3$)	5·6	0·09
Potassium oxide (K$_2$O)	–	0·03
Calcium oxide (CaO)	2·4	–
Magnesium oxide (MgO)	1·2	–
Sodium oxide (Na$_2$O) equivalent	1·7	–
Sulfate (SO$_4$)	0·3	–
Free calcium oxide	0·03	–
Total phosphate	0·5	–
Loss on ignition	< 2·0	0·14
Physical properties		
Size distribution: μm		Retained: %
500	–	0·1
355	–	0·5
250	–	1·5
180	–	6·0
125	–	46·0
90	–	83·0
63	–	96·5

at a fixed dosage rate of 1% by weight of cement. Standard 12 mm long polyvinyl alcohol (PVA) fibres (REC15, Kuraray) were used at a dosage of 2% by volume. The PVA fibres had an average diameter of 39 μm and a tensile strength of 1600 MPa. The surface of the PVA fibres was coated with a proprietary oiling agent (1·2% by weight) to reduce any excessive fibre–matrix chemical bond strength due to their hydrophilic nature.

Sample preparation and curing

A 10-litre Hobart planetary motion mixer was used to prepare all the mixes. Casting was done strictly by placing the materials in one batch to facilitate natural fibre dispersion. All samples were covered with polythene sheeting and then placed in a temperature-controlled laboratory (20 ± 1°C, 53 ± 3% relative humidity). The samples were demoulded after 24 h and then placed in a small curing tank in the same laboratory environment until required for testing. Preloading was conducted on curing day 14.

Data analysis

The UPV was computed using

$$v = \frac{d}{t} \tag{1}$$

where v is the UPV, d is the distance between the two transducers (330 mm) and t is the measured transit time (in μs). The ratio of the pulse velocity recorded in a control sample before and after being placed outdoors (R_c) was computed using

$$R_c = \frac{v_{c,e}}{v_c} \tag{2}$$

where $v_{c,e}$ is the UPV of control samples during natural outdoor exposure and v_c is the UPV of the control samples before being placed outdoors. This parameter gives an indication of the combined influence of continued hydration, moisture and temperature. A similar relationship can be used for preloaded samples

$$R_p = \frac{v_{p,e}}{v_p} \tag{3}$$

in which $v_{p,e}$ is the UPV of the preloaded samples during natural exposure and v_p is the UPV of the preloaded samples before being placed outdoors. The normalised ratio R_n can then be used to determine the extent of self-healing while accounting for the effects of continued hydration

$$R_n = \frac{R_p}{R_c} \tag{4}$$

Test results and discussion

Figure 3(a) shows the average and individual UPV values for the five control samples after being placed outdoors for 82 d. Also presented in this figure are the recorded values of hourly rainfall data; the hourly and daily mean temperature are presented in Figure 3(b). During the test period, the hourly rainfall varied between 0·2 mm and 4·8 mm and the temperature

Figure 3 (a) UPV values for control samples plotted with hourly rainfall data. (b) Variation in temperature

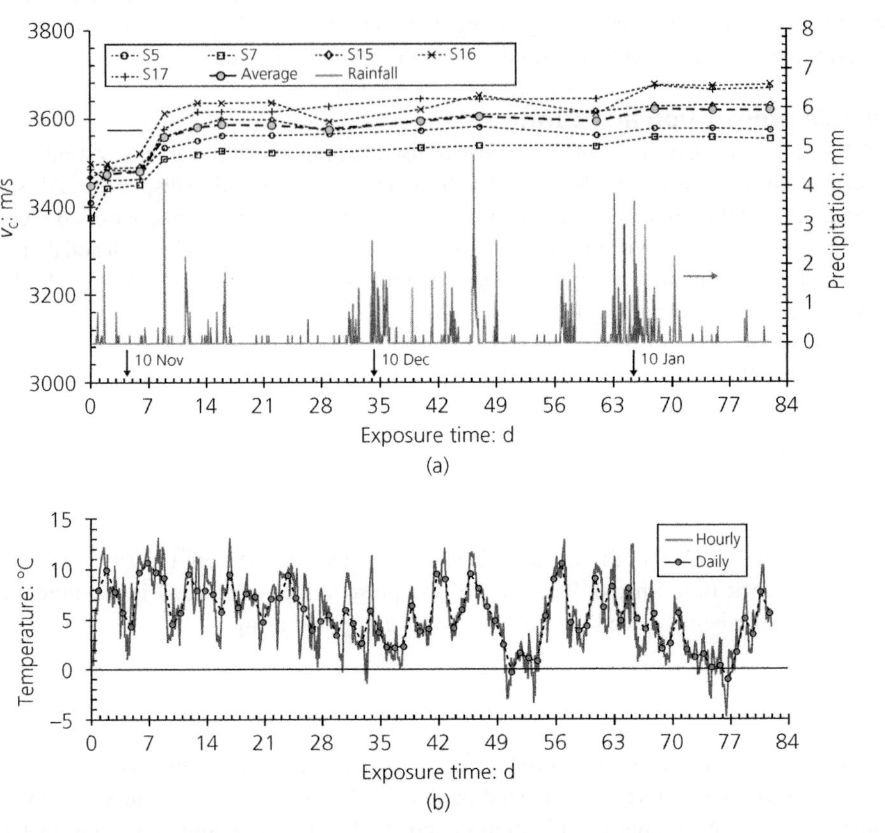

fluctuated between −4·7°C and 13·0°C. It is apparent from the figure that the average UPV values display a general increasing trend, with an average increase of approximately 5% over the test period, from 3446 m/s to 3618 m/s. This increase can be attributed primarily to the change in moisture content within the ECC matrix and progressive refinement of microstructure as a result of ongoing cement hydration and pozzolanic activity. The variations in measured values are likely to be due to variations in moisture content within the ECC and fluctuations in ambient temperature, which affect the rate of hydration. To assess the effect of moisture content, consider, for example, the relatively high increase in UPV values that occurred during the wet period at 9 d of exposure (3556 m/s). Given that the UPV in water is approximately four to five times greater than that in air (Povey, 1997), it is likely that the increase in UPV values during this wet period is attributed to moisture absorbed into the capillary pore network.

Figure 4 presents the normalised UPV values for the preloaded samples calculated using Equation 4. The long-dashed lines, representing samples tested to 60% ultimate tensile strain (samples S4, S6, S13, S14 and S18), display a larger decrease in UPV than the samples tested

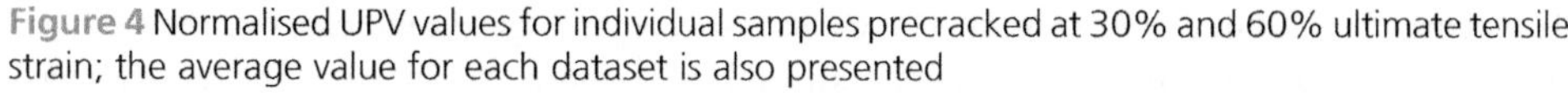

Figure 4 Normalised UPV values for individual samples precracked at 30% and 60% ultimate tensile strain; the average value for each dataset is also presented

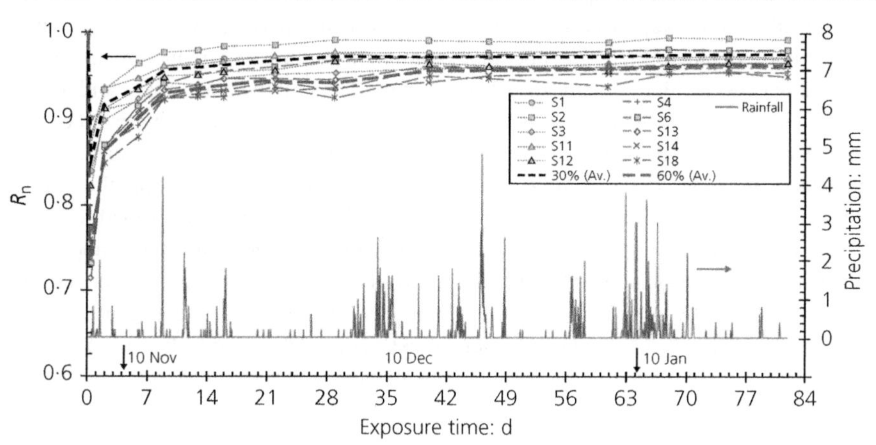

to 30% ultimate tensile strain which are shown as short-dashed lines (samples S1, S2, S3, S11 and S12). In the case of the former, the decrease was approximately 15%, whereas for the latter it was approximately 27%. The decrease in UPV can be attributed to either an increased number of microcracks or an increased width of individual microcracks, or the two effects combined. Based solely on these data, however, it is not possible to determine the relative contribution of each of these factors to the bulk UPV.

Another interesting feature of Figure 4 is the high recovery rate (i.e. slope of the curves) over the first 2 d of outdoor exposure. It is postulated that the high rate of recovery is related to the hydration of unhydrated or partially hydrated cement particles on the surface of narrow microcracks (i.e. crack width $< 10 \, \mu m$). This is the inner core of the cement grains that had previously been encapsulated by hydration products. During cracking, cracks may pass through these unhydrated cement grains and, as water permeates in through the microcracks, the unhydrated cement grains come into direct contact with water, allowing them to hydrate and grow within the space between two cracked surfaces. As these pre-cipitated hydrates infill the microcracks, the signal transit time decreases, thereby leading to the recovery in UPV.

During the first 2 d of outdoor exposure, it is also noticeable that while the more highly damaged samples exhibited a greater loss in UPV, they showed a faster UPV recovery ($\pm 13\%$ recovery compared with $\pm 7\%$ for the less damaged samples). As self-healing depends pri-marily on the initial crack width, with desirable widths less than about $50 \, \mu m$ (Li and Yang, 2007), it can be inferred from the results that it is a larger number of microcracks of $< 50 \, \mu m$ width that is responsible for the initial accelerated healing rate rather than any significant increase in crack width. Should the microcracks have widened significantly, healing would have ceased, leading to a much slower recovery in UPV. This finding also indicates that the larger decrease in UPV exhibited by the more highly damaged samples discussed earlier is primarily due to an increase in the number of microcracks.

Between 2 d and 9 d outdoor exposure, the results presented in Figure 4 indicate that the recovery rate slows down and then plateaus after approximately 9 d of outdoor exposure. The less damaged samples show a UPV value of approximately 96% of that of the control samples, while that of the more damaged samples is approximately 3% lower. During this period, it is postulated that at least two superimposed phenomena are operative within the samples: a reduction in the rate of hydration at narrow-width microcracks as hydration products fill the crack space and the gradual healing of wider microcracks through a combination of the hydration of unhydrated or partially hydrated cement particles and the formation of calcium carbonate on the surface of these microcracks (Kan *et al.*, 2010). From 9 d onward, it is evident from Figure 4 that the self-healing rate within the samples decreased significantly. At 22 d, all samples displayed an almost constant normalised UPV. The normalised UPV for the less damaged samples remained at approximately 97% of that of the control samples, while the equivalent value for the more damaged samples stayed at approximately 94%, indicating a better recovery in the less damaged samples as a result of the presence of a smaller number of microcracks. The relatively constant normalised UPV suggests that the rate of microstructural development in the control samples is comparable to that in the damaged samples (which includes the self-healing process).

It should be noted that the UPV results discussed here provide an indication of physical property recovery and do not necessarily imply any mechanical property recovery. For example, Abdel-Jawad and Haddad (1992) found that the UPV can only detect the occurrence of self-healing and cannot be used to provide an accurate quantification of mechanical strength. By contrast, using the surface wave transmission technique, Aldea *et al.* (2000) and Shah *et al.* (2000) found that it is possible to obtain an indication of mechanical healing of cracks. Since it is the physical properties (i.e. dynamic modulus of elasticity and density) and not the strength per se that influence the transmission of ultrasonic longitudinal pulses through a cement-based material (BSI, 2004), further investigation is required to confirm the recovery in strength of the test samples (i.e. through direct tensile testing).

To gain a qualitative measure of the self-healing process, Figures 5(a), 5(b) and 5(c) show digital micrographs of a damaged area of sample S11 taken before and after being placed outdoors for 6 d and 22 d. As can be seen, there are four fully developed microcracks of different widths in the range 5–50 μm (approximately). Comparing Figures 5(a) and 5(b), it is clear that the initial crack width does affect the existence of self-healing products (detected as white striations on the ECC surface), with microcracks of smaller width showing somewhat denser self-healing products. All microcracks with an initial width of less than 10 μm showed an indication of well-developed healing products, as marked by the formation of white materials. During the same period, microcracks of initial width of approximately 20–30 μm showed a mixed self-healing capability, with the majority showing well-developed healing products (see for example Figures 5(e) and 5(h)). This variation is likely due to the spatial distribution of unhydrated cement grains. These micrograph observations are consistent with the accelerated rate of healing obtained from the UPV measurement during the early stage following a loading event. Partial recovery was seen within microcracks with initial widths of approximately 40–50 μm (see the top of Figure 5(b) for a crack undergoing healing). After 22 d of outdoor exposure (Figures 5(c), 5(f) and 5(i)), it is apparent that all the microcracks were effectively healed. It appears that the white striations are somewhat denser than the surrounding surface on which the presence of a number of occluded air voids is apparent.

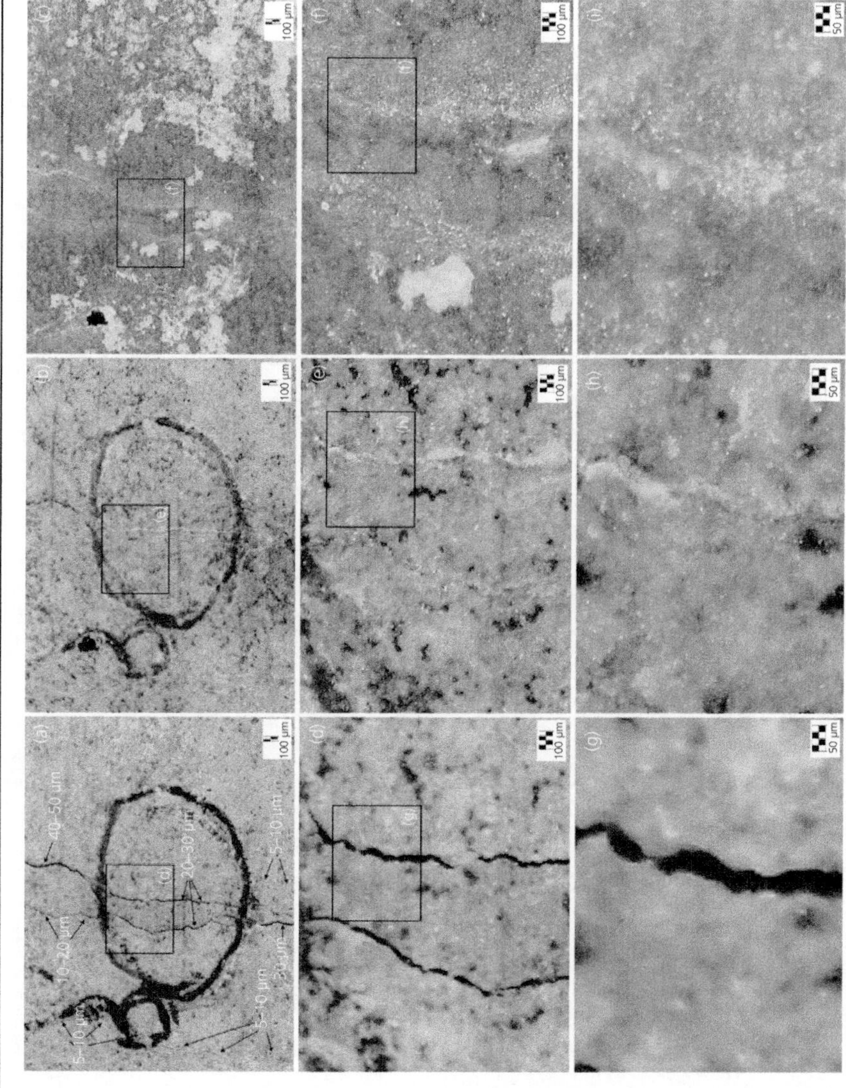

Figure 5 Microcracks observed on the surface of sample S11: (a) before outdoor exposure; (b) after 6 d outdoor exposure; (c) after 22 d outdoor exposure; (d) and (g) are enlarged images of (a); (e) and (h) are enlarged images of (b); (f) and (i) are enlarged images of (c)

Figures 6(a), 6(d) and 6(g) show the largest microcrack (~75 μm) found on the surface of sample S4. There is an apparent growth of self-healing products forming from both crack surfaces into the middle of the crack after 6 d of outdoor exposure (Figures 6(b), 6(e) and 6(h)), with new healing products appearing to build up on top of another. At this time, the crack width has decreased from 75 μm to approximately 25 μm. Figures 6(c), 6(f) and 6(i) show the condition of the crack at 22 d of exposure. As can be seen, most of the space has been filled by

Figure 6 A 75 μm wide microcrack in sample S4: (a) before outdoor exposure; (b) after 6 d outdoor exposure; (c) after 22 d outdoor exposure; (d) and (g) are enlarged images of (a); (e) and (h) are enlarged images of (b); (f) and (i) are enlarged images of (c); (j) after 40 d outdoor exposure; (k) is an enlarged image of (j)

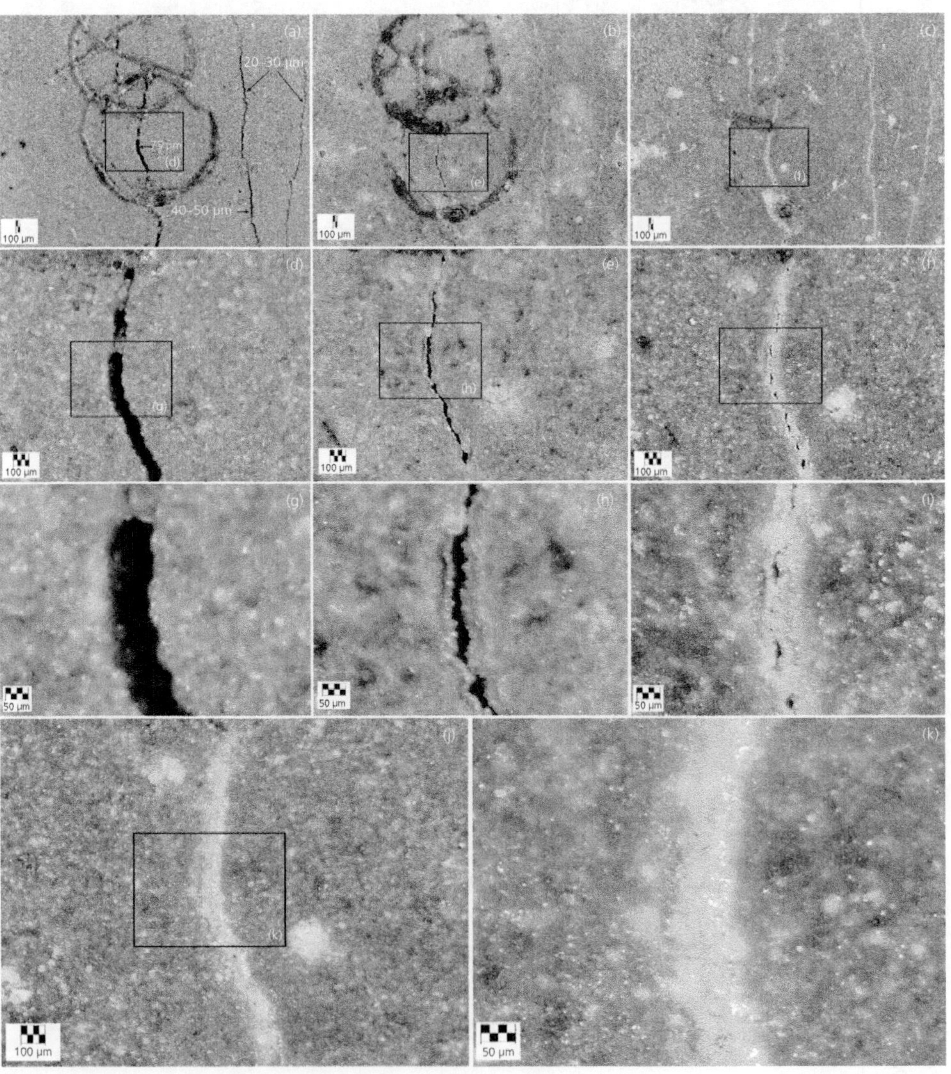

hydration products although some gaps are still visible. As shown in Figures 6(j) and 6(k), the crack was effectively healed after 40 d of exposure.

Conclusions

The following conclusions can be drawn from this study.

- Damage is detected as a decrease in UPV and can be attributed to an increase in the number and width of microcracks. However, a single measurement after preloading does not provide adequate information to allow for determination of the relative contribution of each of these factors to the bulk UPV.
- The control samples displayed a general increasing trend of UPV values over the outdoor test period, highlighting the effects of continuing hydration and pozzolanic activity within the ECC. High values of UPV were found when the samples were wet (i.e. after a rain event).
- All preloaded samples experienced a rapid recovery in UPV over the first 2 d of outdoor exposure, which would imply an intense hydration reaction and possibly the precipitation of calcium carbonate within narrow microcracks (less than about 10 μm).
- The more highly damaged samples displayed a faster recovery rate, although the real values were still lower than those of less damaged samples. The increased recovery rate implies the effective healing of a larger number of microcracks rather than the healing of the same number of wide microcracks. This finding indicates that the larger drop in ultrasonic velocity exhibited by the more badly damaged samples is primarily due to an increased number of microcracks of widths comparable to those seen in the less damaged samples.
- After approximately 9 d of outdoor exposure to intermittent rain, the less damaged samples show a constant UPV value of approximately 96% of that of the control samples, while that of the more damaged samples stayed at approximately 93%.
- The self-healing performance of ECC was shown to be robust. After being exposed to wet weather, all narrow cracks (less than about 10 μm width) healed in less than 6 d, cracks of 20–30 μm width either partially or fully healed in 6 d, 40–50 μm wide cracks fully healed in 22 d, and 75 μm wide cracks partially healed in 22 d and fully healed after 40 d of exposure.

Acknowledgements

The authors acknowledge the support of Kuraray Japan and Kuraray Europe GmbH for providing the PVA fibres and BASF UK for providing the admixtures. Financial support from the School of Energy, Geoscience, Infrastructure and Society is gratefully acknowledged. Thanks also go to Mr D. H. Speranza for assistance in part of the experimental work.

REFERENCES

Abdel-Jawad Y and Haddad R (1992) Effect of early overloading of concrete on strength at later ages. *Cement and Concrete Research* **22(5)**: 927–936.

Aldea BC, Song W, Popovics JS, Member A and Shah SP (2000) Extent of healing of cracked normal strength concrete. *ASCE Journal of Materials in Civil Engineering* **12(1)**: 92–96.

BSI (2004) BS EN 12504-4: Testing concrete – Part 4: Determination of ultrasonic pulse velocity. BSI, London, UK.

Dhir RK, Sangha CM and Munday JGL (1973) Strength and deformation properties of autogenously healed mortars. *ACI Journal Proceedings* **70(3)**: 231–236.

Edvardsen C (1999) Water permeability and autogeneous healing of cracks in concrete. *ACI Materials Journal* **96(4)**: 448–454.

fib (Fédération internationale du béton) (2008) *Bulletin 44: Concrete Structure Management: Guide to Ownership and Good Practice*. fib, Lausanne, Switzerland.

Hearn N (1998) Self-sealing, autogenous healing and continued hydration: what is the difference? *Materials and Structures* **31(8)**: 563–567.

Herbert EN and Li VC (2012) Self-healing of engineered cementitious composites in the natural environment. In *High Performance Fiber Reinforced Cementitious Composite*, vol. 6 (Parra-Montesinos GJ, Reinhardt HW and Naaman AE (eds)). Springer, Dordrecht, The Netherlands; Heidelberg, Germany; London, UK; and New York, USA, pp. 155–162.

Herbert EN and Li VC (2013) Self-healing of microcracks in engineered cementitious composites (ECC) under a natural environment. *Materials* **6(7)**: 2831–2845.

JSCE (Japan Society of Civil Engineers) (2008) *Recommendations for Design and Construction of High Performance Fiber Reinforced Cement Composites with Multiple Fine Cracks (HPFRCC), Concrete Engineering Series 82, Testing Method 6-10*. JSCE, Tokyo, Japan.

Kan LL, Shi HS, Sakulich AR and Li VC (2010) Self-healing characterization of engineered cementitious composites (ECC). *ACI Materials Journal* **107(6)**: 617–624.

Li VC (2003) On engineered cementitious composites (ECC): a review of the material and its applications. *Journal of Advanced Concrete Technology* **1(3)**: 215–230.

Li VC and Yang EH (2007) Self healing in concrete materials. In *Self Healing Materials: An Alternative Approach to 20 Centuries of Materials Science* (Zwaag SVD (ed.)). Springer, Dordrecht, Netherlands, pp. 161–193.

Li VC, Sakulich AR, Reinhardt HW *et al.* (2013) Recovery against mechanical action. In *RILEM State-of-the-Art Reports: Vol. 11 Self-Healing Phenomena in Cement-Based Materials* (de Rooij M, van Tittelboom K, de Belie M and Schlangen E (eds)). Springer, Dordrecht, The Netherlands; Heidelberg, Germany; London, UK; and New York, USA, 3, pp. 119–215.

Long AE, Henderson GD and Montgomery FR (2001) Why assess the properties of near-surface concrete? *Construction and Building Materials* **15(2–3)**: 65–79.

Maekawa K, Ishida T, Chijiwa N and Fujiyama C (2013) Multiscale coupled-hygromechanistic approach to the life-cycle performance assessment of structural concrete. *ASCE Journal of Materials in Civil Engineering* **27(2)**: 1–9.

McCarter WJ, Chrisp TM, Starrs G *et al.* (2010) Developments in monitoring techniques for durability assessment of cover-zone concrete. *Proceedings of 2nd International Conference on Durability of Concrete Structures (ICDCS), Sapporo, Japan*, pp. 137–146.

McCarter WJ, Chrisp TM, Starrs G *et al.* (2012) Developments in performance monitoring of concrete exposed to extreme environments. *ASCE Journal of Infrastructure Systems* **18(3)**: 167–175.

Met Office (2014) *Eastern Scotland: Climate*. See http://www.metoffice.gov.uk/climate/uk/regional-climates/es (accessed 02/12/2014).

Neville A (2012) Autogenous healing – a concrete miracle? *Concrete International* **24(11)**: 76–82.

Pourfalah S and Suryanto B (2013) Development of engineered cementitious composite mixtures using locally available materials in the UK. *Proceedings of the Infrastructure and Environment Scotland 1st Postgraduate Conference, Edinburgh, UK*, pp. 75–78.

Povey MJW (1997) *Ultrasonic Techniques for Fluids Characterization*. Academic Press, New York, NY, USA.

Qian S, Zhou J, de Rooij MR *et al.* (2009) Self-healing behavior of strain hardening cementitious composites incorporating local waste materials. *Cement and Concrete Research* **31(9)**: 613–621.

Qian S, Zhou J and Schlangen E (2010) Influence of curing condition and precracking time on the self-healing behavior of engineered cementitious composites. *Cement and Concrete Research* **32(9)**: 686–693.

Reinhardt HW and Jooss M (2002) Permeability and self-healing of cracked concrete as a function of temperature and crack width. *Cement and Concrete Research* **33(7)**: 981–985.

Reinhardt HW, Jonkers H, van Tittelboom K *et al.* (2013) Recovery against environmental action. In *RILEM State-of-the-Art Reports: Vol. 11 Self-Healing Phenomena in Cement-Based Materials* (de Rooij M, van Tittelboom K, de Belie M and Schlangen E (eds)). Springer, Dordrecht, The Netherlands; Heidelberg, Germany; London, UK; and New York, USA, 3, pp. 65–117.

Shah SP, Popovics SJ, Subramaniam KV and Aldea CM (2000) New directions in concrete health monitoring technology. *Journal of Engineering Mechanics* **126(7)**: 754–760.

Sisomphon K, Copuroglu O and Koenders EAB (2013) Effect of exposure conditions on self-healing behavior of strain hardening cementitious composites incorporating various cementitious materials. *Construction and Building Materials* **42**: 217–224.

Snoeck D, van Tittelboom K, Steuperaert S, Dubruel P and de Belie N (2014) Self-healing cementitious materials by the combination of microfibres and superabsorbent polymers. *Journal of Intelligent Material Systems and Structures* **25(1)**: 13–24.

Suryanto B, Nagai K and Maekawa K (2010) Bidirectional multiple cracking tests on high-performance fiber-reinforced cementitious composite plates. *ACI Materials Journal* **107(5)**: 450–460.

Vaysburd AM, Brown CD, Emmons PH and Bissonnette B (2005) Some thoughts on realcrete, labcrete, and designcrete in concrete repair. *Concrete Repair Bulletin* **July/August**: 24–29.

WeatherCast (2014) *Weather Reports Worldwide*. See http://www.weathercast.co.uk/world-weather/weather-stations/obsid/3166.html (accessed 05/11/2014).

Yang Y, Lepech MD, Yang EH and Li VC (2009) Autogenous healing of engineered cementitious composites under wet–dry cycles. *Cement and Concrete Research* **39(5)**: 382–390.

Yang Y, Yang EH and Li VC (2011) Autogenous healing of engineered cementitious composites at early age. *Cement and Concrete Research* **41(2)**: 176–183.

Dhir and Paine
ISBN 978-0-7277-6457-7
https://doi.org/10.1680/icetsc.64577.085
ICE Publishing: All rights reserved

Chapter 6

Biomimetic cementitious construction materials for next-generation infrastructure

Abir Al-Tabbaa
Professor of Geotechnical Engineering, University of Cambridge, Cambridge, UK
(corresponding author: aa22@cam.ac.uk)

Bob Lark
Emeritus Professor of Civil Engineering, Cardiff University, Cardiff, UK

Kevin Paine
Reader in Civil Engineering, University of Bath, Bath, UK

Tony Jefferson
Professor of Civil Engineering, Cardiff University, Cardiff, UK

Chrysoula Litina
Research Associate in Resilient Materials for Life Programme Grant, University of
Cambridge, Cambridge, UK

Diane Gardner
Senior Lecturer in Civil Engineering, Cardiff University, Cardiff, UK

Tim Embley
Group Innovation and Knowledge Manager, Costain Ltd, Maidenhead, UK

The resilience of civil engineering structures has traditionally been associated with the design of individual elements with sufficient capacity to respond appropriately to adverse events. This has traditionally employed 'robust' design procedures that focus on defining safety factors for individual adverse events and providing redundancy. As such, construction materials have traditionally been designed to specific technical specifications. Furthermore, material degradation is viewed as inevitable and mitigation necessitates expensive maintenance regimes. Based on a better understanding of natural biological systems, biomimetic materials that have the ability to adapt and respond to their environment have recently been developed. This fundamental change has the potential to facilitate the creation of a wide range of 'smart' materials and intelligent structures that can self-sense and self-repair without the need for external intervention which could transform infrastructure. This paper presents an overview of the development, application and commercial perspectives of a suite of complementary self-healing cementitious systems that have been developed as part of a national team and led to the first UK full-scale field trials on self-healing concrete.

1.   Introduction

The UK's current ageing and deteriorating infrastructure is the result of decades of underinvestment that is now costing the nation £2 million/d (Kelly, 2014) and has resulted in the UK government heavily investing in both existing and new infrastructure, committing ~£500 billion by 2020–2021 (HMT, 2016). Current figures show that half of the construction budget is being spent on the repair and maintenance of (mainly concrete) infrastructure at £40 billion/year (ONS, 2016). Moreover, the durability of repaired concrete structures continues to be a major concern as after 5 years, 20% of the repairs fail, increasing to 55% after 10 years (Tilly and Jacobs, 2007). Furthermore, these figures are compounded when the cost of disruptions and delays is considered, which could be up to ten-times that figure (Highways England, personal communication, 2018).

A major challenge is that civil engineering design practices remain traditional, based on the assignment of appropriate partial material and action factors and the provision of a degree of redundancy to prevent failure. Hence, construction materials are designed to meet a prescribed specification: material degradation is viewed as inevitable and mitigation necessitates expensive inspection, maintenance, repair and eventually replacement regimes. It is unsurprising, therefore, that poor material performance is the single main cause of deterioration and failure in infrastructure systems worldwide (Tilly and Jacobs, 2007).

These durability problems would be greatly diminished if construction materials could self-heal or mimic natural biological materials. These 'biomimetic' materials are advanced materials that can transform infrastructure by embedding resilience within its components and systems so that rather than being defined by individual events, they can evolve and adapt over their lifespan. Recent national and international governments and industry road mapping reports have highlighted that advanced infrastructure materials, with specific reference to biomimetic attributes, will play a pivotal role in the transition of infrastructure into a low-carbon dioxide and resilient future (EC, 2014; GO-Science, 2017; i3P, 2017; WEF and BCG, 2016). This transition will necessitate a step change in the value placed on infrastructure materials and provide the much higher level of confidence needed in the reliability of the performance of infrastructure systems.

Most concrete infrastructure assets are regularly exposed to a myriad of damaging environmental actions, resulting in different forms of damage such as cracking, weakened areas of material, oxidation of reinforcement, loss of rebar cover, material dissolution, swelling and expansion. To date, the most extensively studied damage mode in cementitious systems, concrete being one example, is cracking. In this respect, self-healing phenomena in cementitious systems are broadly divided into two categories: autogenic and autonomic. Autogenic self-healing refers to self-healing processes that are an intrinsic characteristic of the components of the cementitious matrix, effective for small crack widths of ≤ 0.15 mm. In contrast, autonomic self-healing refers to actions that use components that do not naturally exist in the cementitious matrix – that is, 'engineered' additions – usually employed to deal with larger crack sizes. It is worth pointing out that self-healing phenomena would apply to all forms of damage listed earlier, but the focus of this paper is on cracking. A range of self-healing cementitious systems have been investigated by the research community and examples are shown schematically in Figure 1. For a detailed overview of the self-healing phenomena shown in Figure 1, see the publications by de Rooij *et al.* (2013) and De Belie *et al.* (2018). Different autogenic self-healing systems do work in a synergistic way. Similarly, autogenic and autonomic self-healing systems can potentially work in combination such that the autonomic system reduces the crack size and enables autogenic processes to complete the self-healing process. Ultimately, autonomic systems could be designed to work in tandem, which forms the basis of the work initiated in this paper.

Figure 1 Self-healing phenomena in cementitious systems: (a) autogenic self-healing (figure adopted from de Rooij *et al.* (2013)) and (b) autonomic self-healing (figure adopted from Souza (2017). SAP, superabsorbent polymers

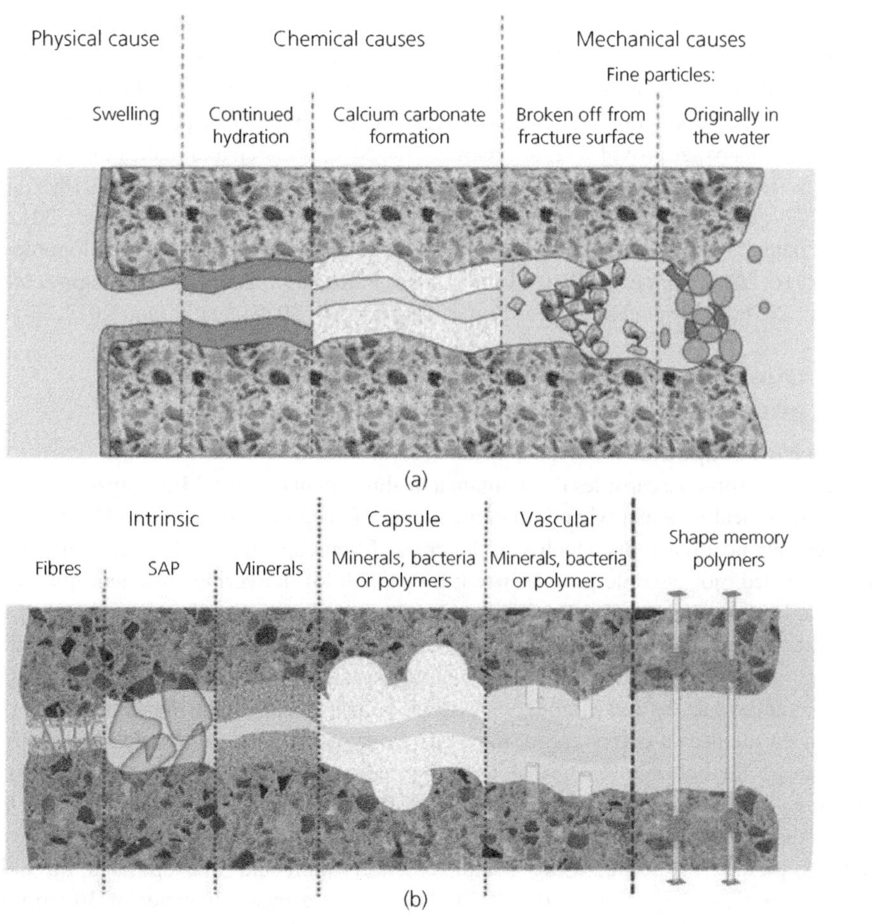

The research presented in this paper was undertaken between 2013 and 2016 (Al-Tabbaa *et al.*, 2017; Lark *et al.*, 2013; Paine *et al.*, 2015) as part of the Engineering and Physical Sciences Research Council-funded Materials for Life (M4L) project with the objectives of (*a*) developing a suite of complementary autonomic self-healing systems that can act for synergistic application to address physical damage (cracks) at a range of scales and (*b*) demonstrating and implementing these self-healing systems in real applications. In particular, the research presented in this paper is focused on the work of the M4L research team to develop four self-healing systems: microcapsules, calcite-precipitating bacteria, shape memory polymers (SMPs) and vascular networks (Figure 1(b)). The specific success and innovations in each case included the development of (*a*) microcapsules with switchable mechanical properties and with mineral healing cargo; (*b*) bespoke combinations and microencapsulation of alkaliphilic bacteria, nutrients and precursors for rapid precipitation of healing products; (*c*) remotely activated SMP multistrand sheathed tendons; and (*d*) innovatively created vascular networks using polyurethane tubing.

Subsequent sections of the paper provide a brief summary of the experimental laboratory work carried out to develop each of these four self-healing systems, including a summary of key findings, details of field tests undertaken in 2015 to validate the scaling up of those self-healing systems and to validate their performance using the A465 Highway development project as a case study. The paper concludes with an overview of a number of relevant commercial perspectives. This paper provides an overview of the M4L research and makes explicit reference to a number of other published outputs produced by M4L (Alazhari *et al.*, 2018; Al-Tabbaa *et al.*, 2017; Davies *et al.*, 2015, 2016, 2018; De Belie *et al.*, 2017; Gardner *et al.*, 2012, 2014, 2017; Giannaros *et al.*, 2016; Isaacs *et al.*, 2013; Jefferson *et al.*, 2010; Kanellopoulos *et al.*, 2015, 2016, 2017; Lark *et al.*, 2013; Litina, 2015; Paine *et al.*, 2015, 2016; Pilegis *et al.*, 2015; Qureshi, 2016; Sharma *et al.*, 2017; Souza, 2017; Teall, 2017; Teall *et al.*, 2017, 2018).

2. Overview of the developed self-healing systems and associated laboratory work

2.1 Microcapsules

These are micron-size capsules that contain a healing agent (cargo). Microcapsules rupture and release their healing agent when in the direct path of an approaching crack. The healing agent initiates chemical reactions which seal the crack. Microcapsules must rupture only when and where needed, not degrade prior to need; be capable of delivering sufficient quantity of an effective healing agent; and not undermine the structural integrity of the cementitious system. Two microencapsulation systems – namely, emulsification polymerisation and microfluidics – were set up to produce laboratory-scale quantities of microcapsules that enabled the production of a range of shells and cargo (Litina, 2015; Souza, 2017). While most research to date has focused on the use of epoxy cargo, the M4L work focused on the development of mineral-based cargo for better compatibility with the cementitious matrix and better shelf-life and handling. The most promising developments included microcapsules with polymeric gelatin/gum arabic shell and sodium silicate cargo (see Figure 2(a)), which was developed with an industry partner, Lambson, using complex coacervation and was upscaled for field trial applications (Kanellopoulos *et al.*, 2017). These have a mean diameter of 300 μm and are ductile, to survive the mixing initially, and then brittle within the hardened cement, to rupture under crack propagation.

Figure 2 Self-healing microcapsules: (a) the microcapsules developed with Lambson for the field trials (Kanellopoulos *et al.*, 2017); (b) the microcapsules embedded in cement (Giannaros, 2017); (c) a ruptured microcapsule with healing products (Kanellopoulos *et al.*, 2017); (d) an example of a healing product scaffold (Qureshi, 2016); (e) crack healing in action (reproduced from Kanellopoulos *et al.* (2015), by kind permission of Elsevier)

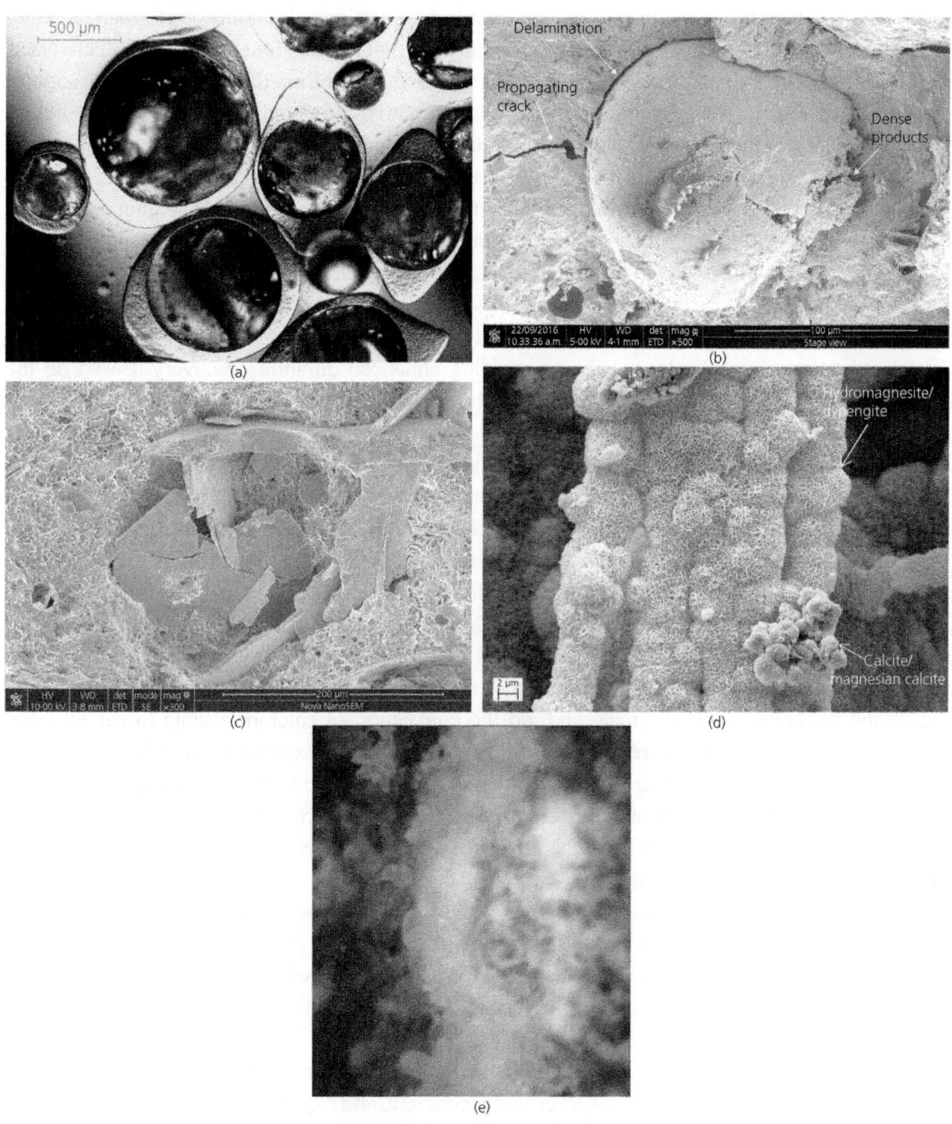

The effective mixing and survivability of the microcapsules within the cementitious matrix was demonstrated with microstructural imaging (see Figure 2(b)). Extensive laboratory investigations were then carried out to investigate the effect of the polymeric microcapsules on the fresh and hardened properties of cementitious matrices (cement, mortar and/or concrete) (Kanellopoulos *et al.*, 2016). Microcapsule dosages of up to 32% by volume (10·7% wt) were tested. Results showed that up to 8% by volume (~2·7% wt) addition of these microcapsules to the cement had negligible detrimental effect on the workability and mechanical properties of the cementitious matrix. Optimisation studies in terms of the self-healing efficiency under different damage scenarios and underwater curing (Giannaros *et al.*, 2016; Kanellopoulos *et al.*, 2015) were also undertaken. Crack widths between 0·11 and 0·25 mm were tested. Microstructural analyses, to understand the healing mechanisms, and mechanical testing and durability performance, using sorptivity and permeability, were carried out, as the common metrics for self-healing effectiveness (Ferrara *et al.*, 2018). Scanning electron microscopy imaging confirmed the release of the sodium silicate cargo, which reacted with the cementitious matrix, producing an excess of hydration products (illustrated in Figure 2(c)), which sometimes formed a stable scaffold for the proliferation of carbonate crystals that deposited due to autogenic healing (illustrated in Figure 2(d)). This illustrated a new mechanism through which autonomic and autogenic self-healing interacted. Complete healing was observed after 28 d (illustrated in Figure 2(e)). A strong correlation between healing potential and microcapsule content was identified. Increased strength recovery compared to the control, with no microcapsules, (~25%), was observed. A reduction in sorptivity values compared to the control (~50%) and a permeability reduction of up to one order of magnitude were critical indicators of self-healing success providing enhanced durability recovery. Based on the laboratory work, a dosage of 8% by volume of the cement was selected for the field trials.

2.2   Calcite-precipitating bacteria

The general principle here is that bacteria are encapsulated in the cementitious matrix and, provided that there is an appropriate source of calcium available, they can precipitate calcium carbonate in cracks under favourable conditions. There are a number of key pathways dependent on the bacteria used (De Belie *et al.*, 2017): (*a*) enzymatic hydrolysis of urea, (*b*) dissimilation of nitrates and (*c*) aerobic metabolic conversion. Research within M4L concentrated on the last approach, in which spores of the bacteria were encapsulated in concrete together with calcium acetate and nutrients at the mixing stage. When cracks occur in the concrete, ingress of water and oxygen then occurs and the spores can germinate and aid the conversion of calcium acetate to calcium carbonate. Carbon dioxide is also released, and this results in further carbonation of hydrates in the locality of the crack, promoting further healing. For self-healing to work, an appropriate non-ureolytic species capable of germinating and precipitating calcium carbonate sufficiently quickly in the alkaline conditions present in concrete is required. Based on literature, three species of bacteria were selected for detailed investigation as described elsewhere (Sharma *et al.*, 2017). This led to the selection of *Bacillus pseudofirmus* DSM 8715 (Figure 3(a)) for use in further studies. It produces spores readily in short time frames (24 h), and its spores germinate rapidly and achieve almost full germination within 2 h (Sharma *et al.*, 2017). Further tests showed that the survivability of the spores in cement paste was low and on the order of only 1–4% after 93 d.

Because problems with survival of the spores may have been a limiting factor, it was considered necessary to take steps to protect the spores in the concrete, and expanded perlite was used for this purpose. The growth medium (calcium acetate and yeast extract) was also encapsulated in expanded perlite, but, unlike other research (De Belie *et al.*, 2018), the spores and growth medium were not encapsulated in the same particles. It was found that self-healing could be observed visually provided that there was an appropriate ratio of spores to growth media (illustrated in Figures 3(b) and 3(c)). Self-healing was quantified by means of a modified initial surface absorption test (bespoke), and it was shown that those specimens in which visual healing could be

Figure 3 Bacterial healing: (a) spores of *B. pseudofirmus*; (b, c) self-healing of a crack by the action of *B. pseudofirmus*

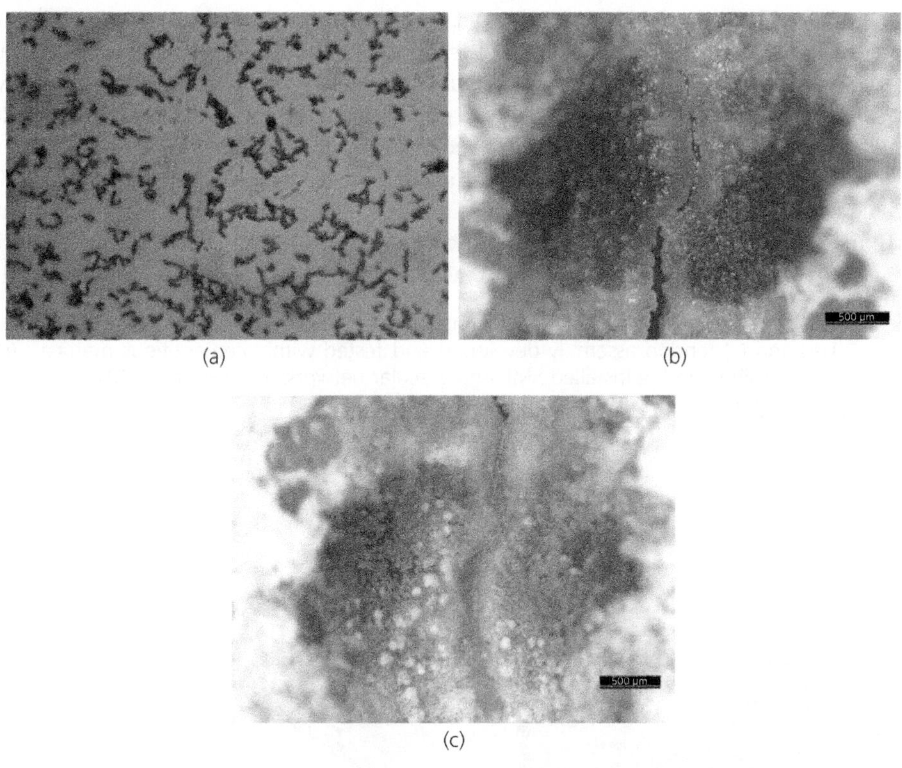

observed had the lowest initial surface absorption values as expected. Parallel microbiological tests reported by Alazhari *et al.* (2018) have suggested that it may be possible to achieve greater healing with the use of a multicomponent growth medium consisting of ingredients to promote germination of spores and growth of the bacteria further, although, for the full-scale field trials, only yeast extract was utilised as a source of nutrients. Prior to the field trials, the M4L team also undertook further laboratory trials to demonstrate that concrete with encapsulated calcite-precipitating bacteria met appropriate technical specifications for consistency and strength (Paine *et al.*, 2016).

2.3 Shape memory polymers

SMPs are materials that respond to external stimuli by altering their physical shape. A novel approach of incorporating heat-activated SMPs into cementitious systems allows post-construction cracks to be closed. There is also the added benefit that the reduction in the size of the cracks facilitates autogenic self-healing (Isaacs *et al.*, 2013). Thus, these tendons form part of a self-healing system in which the tendons are activated when damage is detected. In M4L, tendons formed from amorphous SMP filaments, formed by die drawing, were heated above their glass transition temperature (T_g) while restrained. This process generates a restrained shrinkage stress that serves to close cracks. The restrained tendons also apply a small compressive stress across the crack faces, which aids autogenic healing. While earlier studies had demonstrated the feasibility of SMPs using commercially available polyethylene terephthalate (Pet) (Isaacs *et al.*, 2013; Jefferson *et al.*, 2010), findings from the M4L work (Isaacs *et al.*, 2013; Jefferson *et al.*, 2010) showed that SMPs reduce crack widths (0·5 mm wide) by an average 74%. This reduction rose to 93% when specimens were healed in water. In addition to these results, M4L work found that autogenic self-healing made possible by the autonomic use of SMPs produced a 13% load recovery (defined as the ratio of the maximum load achieved after healing to the maximum load achieved prior to the creation of the damage) after 28 d healing, in comparison with 1% for control specimens. For this technique to be employed in the field, it was necessary to develop a new tendon assembly and a remote activation technique. The technique developed by M4L for this purpose used a tendon comprising multiple 0·9 mm dia. SMP filaments, an outer spiral wire for electrical activation and a plastic sheath for protection (Figures 4(a) and 4(b)) (Pilegis *et al.*, 2015; Teall *et al.*, 2017). As reported by Teall *et al.* (2018),

Figure 4 SMPs for self-healing: (a) a typical example of an SMP shape (Teall, 2017); (b) SMP (Davies *et al.*, 2018) and (c) tendon assembly developed and tested within cementitious matrices; (d) a section of a wall showing the installed SMP and vascular networks (Davies *et al.*, 2018)

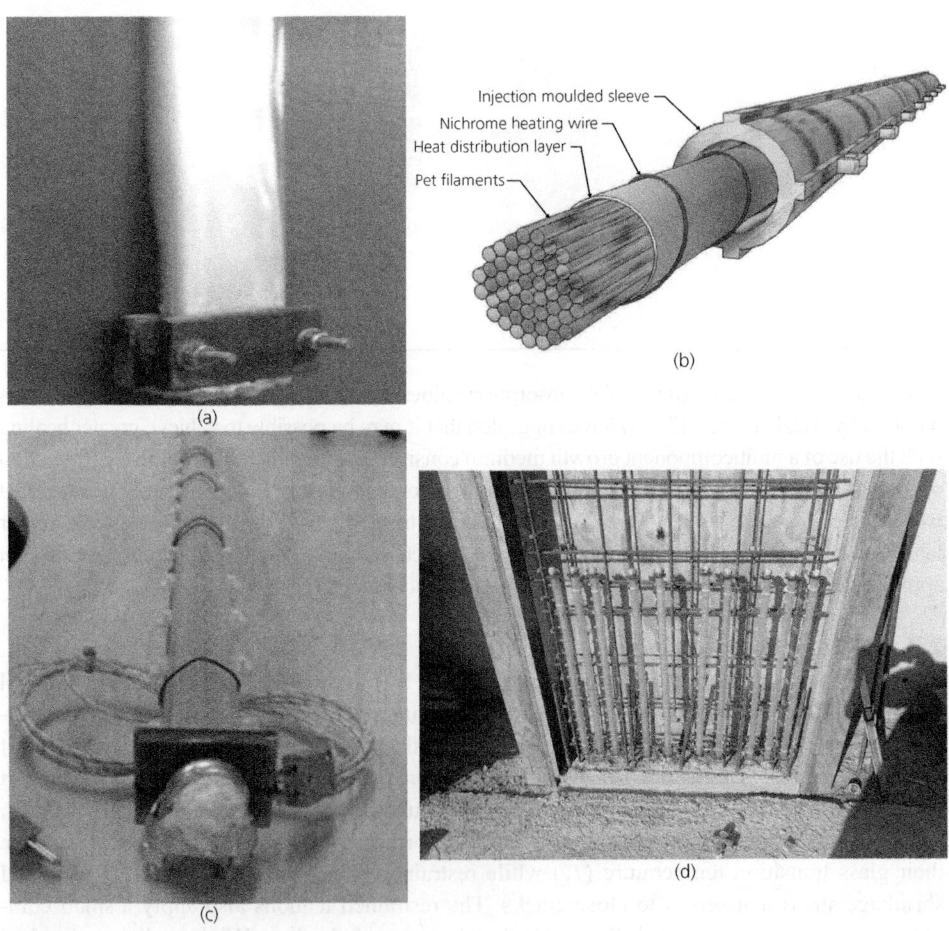

these were further tested in $500 \times 100 \times 100\,\text{mm}^3$ concrete beams, but this time half of the specimens also included two 6 mm dia. high tensile ribbed steel bars placed either side of the SMP tendon to replicate the restraint that will be provided by traditional rebar. From this study (Teall *et al.*, 2018), it was concluded that restrained shrinkage stresses of 20 MPa could be generated in the tendon, which resulted in crack closures of up to 85% and a significant increase in beam stiffness on reloading in unreinforced beams. In contrast, in the reinforced beams, crack closure was only just over 30%, which suggests that a higher shrinkage stress is needed for the system to be truly effective in reinforced-concrete elements.

2.4   Vascular networks

To address the additional challenge of repeat self-healing, M4L research investigated the benefits of incorporating vascular networks in concrete elements to deliver healing agents to cracked regions. These networks comprise connected capillary channels as illustrated in Figures 5(a)–5(c). A new method for creating interconnected channels within a concrete structural element was developed. The method involves placing interconnected polyurethane tubes in a mould or shutter prior to casting and subsequently withdrawing them when the

Figure 5 Vascular networks in (a) laboratory experiments, (b) large-scale laboratory testing and (c) the field trials' wall panel

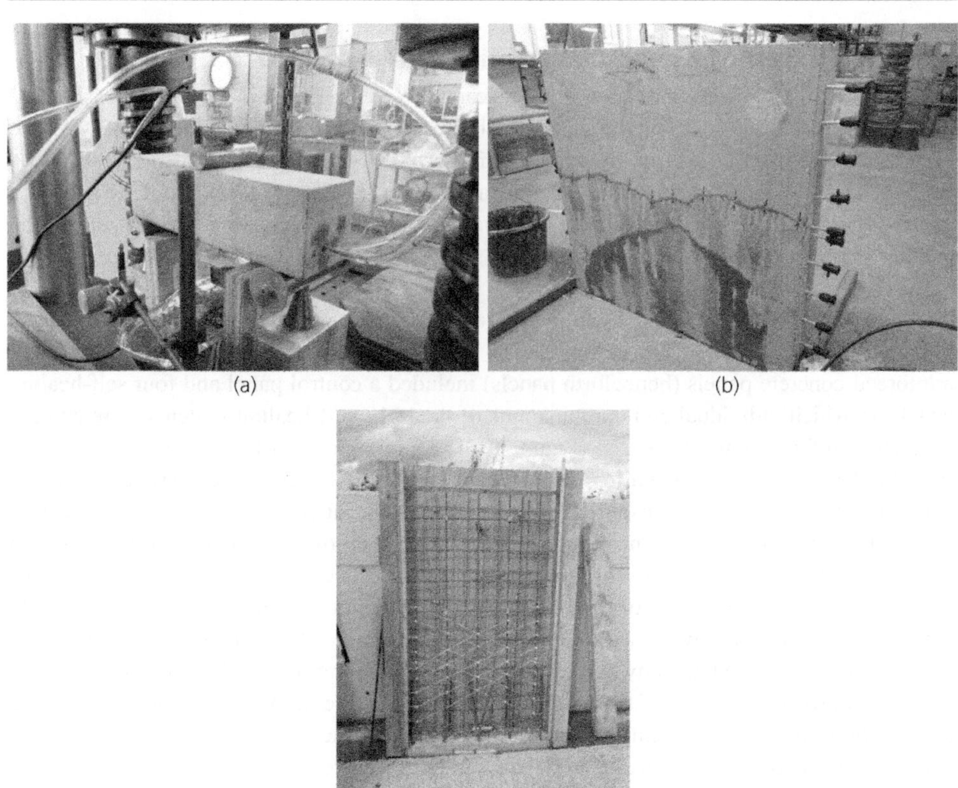

(a)

(b)

(c)

concrete has hardened, thereby leaving a permanent network of channels (or vascular networks) within the body of the concrete member (Davies *et al.*, 2015). This method was shown to be sufficiently robust and scalable to be used on a construction site. To facilitate the use of vascular networks, the M4L team designed and undertook a series of combined numerical–experimental studies to investigate the capillary flow properties of healing agents in concrete cracks. These studies explored the transport behaviour of a range of natural and autonomic agents in manufactured (smooth) and natural (rough) cracks of varying apertures (Gardner *et al.*, 2012, 2014, 2017). This research established a series of parameters that characterise viscous resistance, capillary tension, contact angle, meniscus resistance, stick-slip and wall resistance. In addition, based on these studies, a model that uses the Lucas–Washburn equation for capillary flow for accurately predicting the flow properties of a range of healing agents in cracks was developed and validated by the M4L team (Gardner *et al.*, 2017). Based on the findings of these characterisation studies, cyanoacrylate, commonly known as 'superglue', was chosen for use in the vascular network laboratory tests because of its low viscosity and rapid polymerisation. However, for site trials, for which a longer healing time was preferable, sodium silicate was chosen as the healing agent. The laboratory test series showed that healing agents can be delivered by capillary flow alone but that it is beneficial to apply a small additional pressure to improve healing agent delivery. For this reason, a pressure of 0·1 bar above atmospheric pressure was applied to the healing agent in the site trials.

3.   Field deployment and applications to date

Findings from the preceding laboratory research to deliver four autonomic self-healing systems (microcapsules, calcite-precipitating bacteria, shape memory tendons and vascular networks) were used to inform the design of full-scale field trials as part of the Welsh government's A465 Heads of the Valleys Highways Upgrade scheme, for which Costain is the main contractor. The purpose of the full-scale field trials was to assess the full-scale application and performance of the developed autonomic self-healing systems in the field.

The field trials involved the construction of five full-scale concrete retaining wall panels, each 1·8 m high and 1 m wide (Figure 6) (Davies *et al.*, 2016, 2018; Teall *et al.*, 2016). These reinforced-concrete panels (henceforth panels) included a control panel and four self-healing panels, in which individual and combinations of the M4L self-healing systems were embedded. The reinforcement was placed (Figure 6(a)) to allow cracking to occur at around 0·5 m from the base of the wall when loaded as a cantilever. The concrete used throughout was a ready-mix C40/50 concrete, using CEM I and limestone aggregates and with a water-to-cement ratio of 0·43. The self-healing systems used in each of the panels are summarised in Table 1. Panels A and C were used to investigate the influence of microcapsules and bacteria, respectively, on the behaviour of the concrete and their ability to provide autonomic healing when damaged. Panel B was used to assess the ability of the SMP tendons to both close the cracks that were formed and any subsequent autogenic healing. Panels D and E were used as controls to provide a record of the behaviour of a standard reinforced-concrete wall without any additions and one containing an empty embedded flow network. At a later date, the flow networks in panels B and E were flushed with a sodium silicate solution to assess the ability of the networks to deliver a healing agent and whether the degree of healing was enhanced by the use of the tendons to close the cracks.

Figure 6 The M4L field trials of self-healing concrete wall panels at the A465 site (a) during construction and (b) after demoulding. Visible from left to right in both (a) and (b) are panels A, B,C, D and E. Details of self-healing systems are available in Table 1

(a)

(b)

Table 1 Trial panels and their embedded self-healing mechanisms (Davies *et al.*, 2018)

Panel	Self-healing system/materials used in the panel
A	Microcapsules containing sodium silicate cargo
B	SMP tendons and an empty embedded flow network
C	Bacteria and nutrient-infused perlite and an empty embedded flow network
D	Control
E	Control with an empty embedded flow network

Based on the results from M4L studies on microcapsules, panel A was constructed using a concrete mix with microcapsules at 8% by volume of cement (~2·7% by weight), which corresponded to 0·47% by weight of the total concrete mix. The microcapsules, supplied in a preserving solution, were first washed with water and filtered before being added, in their slurry form, directly into the concrete mix using a portable 120L Belle concrete mixer. In panel B, vascular networks were combined with SMP tendons. A two-dimensional network of 4 mm dia. channels was created using polyurethane tubes, which were removed from the concrete after the formwork had been struck. The channels were connected using three-dimensional (3D)-printed joints made from poly(lactic acid), which were tied to the outermost reinforcement. At either side of the panel, the flow networks were terminated with lockable steel injection packers, which allowed each channel to be sealed individually to aid the loading and pressurising of the network. Sodium silicate solution was pumped through the tubes using a nominal pressure of 0·1 bar. This panel also included ten 750 mm long SMP tendons, which were designed, upon activation, to generate a stress of 0·5 MPa across the cracked external face. Each tendon contained 200 Pet filaments that were surrounded by a heating system and a 3D-printed polymer sleeve. The Pet filaments, manufactured by Bradford University, were capable of generating a restrained shrinkage stress of 30 MPa in the laboratory. These tendons were placed at an eccentricity of 40 mm in a staggered layout. In panel C, the components for bacterial healing were encapsulated before addition to the concrete mix and only used in the zone surrounding the expected location of the main crack. Coated perlite impregnated with spores was used at a content of 45 kg/m^3, equating to ~4 × 10^{13} spores/m^3 of concrete, while the coated perlite containing nutrients was included at 140 kg/m^3 and provided 3·8% calcium acetate by mass of cement and 0·1% yeast extract by mass of cement to the concrete. In addition, two panels (panels D and E) were cast which acted as controls. One was a standard reinforced-concrete panel, and the second panel had an empty flow network with no healing agent. Both panels were used for comparison purposes.

The panels were constructed in September 2015 and then subjected to damage and allowed to subsequently heal. The damage the panels were subjected to was a simulation of that they would experience under normal use. The responses of the panels in both pre- and post-healed conditions were monitored with a range of in situ instrumentation and visualisation techniques. Both the scaling-up process and the field trial applications successfully demonstrated the feasibility of commercial production and application of those self-healing systems. All self-healing panels exhibited measurable healing responses, in terms of both stiffness/strength recovery and crack filling. The field trials also highlighted a number of challenges and areas for further improvement in the self-healing systems, and further research has been initiated in response to these challenges. Although not all SMP tendons in panel B were successfully activated, an average reduction in crack width of ~20% and an increase in stiffness upon reloading of the panel were achieved. Furthermore, the vascular network in the same panel was shown by visual surface observations to be able to deliver healing agent to fill cracks but produced only limited mechanical healing. The microcapsule panel (panel A) confirmed the real-time feasibility of microcapsule-based healing, showing improved crack-size reduction (width and depth) of ~60% as early as 1 month and recovery in permeability (by ~2·5 orders of magnitude) at 6 months of self-healing. While some healing was identified in the section of panel C that contained the bacteria, the extent of self-healing was less than anticipated. Based on the working hypothesis that the lower than anticipated level of self-healing was caused by cold weather, further work has now been initiated by the research team to identify bacteria more appropriate for use in low temperatures. While it is acknowledged that the field conditions were far more severe than the laboratory-imposed conditions, the results from the field trials are extremely promising.

4. Discussion

4.1 Potential commercial benefits

There has been significant interest to date from companies across the whole supply chain, contractors, consultants, material suppliers and construction chemical companies, and from client organisations, who have identified practical applications for the four self-healing systems. The huge appetite for self-healing cementitious systems of the construction industry and their potential benefits is captured and documented in market research conducted for M4L by Lychgate in 2016, who surveyed and interviewed over 40 major construction organisations (Gardner *et al.*, 2018). The survey provided evidence of the continuing problems with current cementitious materials in existing and new concrete structures and identified potential applications for self-healing cementitious systems in bridges and highways, marine and water-retaining structures, tunnels (at joints, in sprayed concrete and precast elements), nuclear installations, dams, concrete piles and airport runways. The benefits anticipated by survey respondents include reduced use and costs of overdesign to provide resilience, the need for fewer additives and reduced concrete cover and reinforcement.

The significant advances and recent successes made in the Netherlands on the development of bacteria-based self-healing for concrete (referred to as bioconcrete), currently commercialised through the spin-out company Basilisk, provides confidence in the proposed technologies. Furthermore, the development and commercial application of self-healing asphalt, with induction heating, also in the Netherlands, illustrates the type of benefits that can similarly be realised through the application of the self-healing systems described in this paper to concrete. The first self-healing asphalt road in the Netherlands was constructed in 2010 on a 400 m stretch on the A58 motorway near Vlissingen and was donated by the government as a test bed. The motorway has been in public use since. This and similar application on 12 other Dutch roads, all of which are in a good condition, have led to conclusions that the life of these asphalt roads can be doubled if they are melted every 4 years and demonstrated that self-healing asphalt could save the Dutch €90 million/year in routine maintenance costs, if used on all Dutch roads (Interesting Engineering, 2017).

4.2 Impact on whole-life costs

The M4L project undertook various cost and economic feasibility analyses and studies. In the Lychgate market research study outlined earlier (Gardner *et al.*, 2018), a 20% increase in the initial cost was considered acceptable by concrete producers, although it was clear that they would be aiming for a much lower premium. A whole-life costing (WLC) exercise for a bridge deck under construction as part of the A465 Heads of the Valleys road improvement project was carried out as a case study (Teall, 2017). Adaptations were made to deterioration curves in the Structures Asset Management Planning Toolkit to allow self-healing concrete deck slabs to be modelled over varying numbers of self-healing cycles. Comparisons of the costs associated with a conventional reinforced-concrete deck slab against a self-healing concrete deck slab capable of one, two, three and *n* (indefinite) healing cycles, were performed (Teall, 2017).

The results demonstrated that a concrete deck slab capable of only a single healing cycle would not reduce the WLC of the structure. A slab capable of multiple healing cycles could enable a WLC saving of up to ~12%. Given the relatively low material cost of the concrete deck slab compared to the potential savings over the 120-year study period, this finding means an increase in material cost of up to 20 times could be justified for the use of a self-healing concrete capable of multiple healing cycles. A sensitivity analysis also highlighted that a self-healing concrete deck slab is most viable for structures subject to severe exposure conditions.

4.3   Main challenges and barriers

The main challenges and barriers which need to be addressed to enable the maximum uptake of the technology are the same as those associated with the introduction of any new construction materials. Specifically, these self-healing concrete technologies include the need for (*a*) extensive validation data to provide confidence, (*b*) material certification to prove compliance with national and international standards, (*c*) design procedures and analysis methods to be developed before the proposed systems can be safely applied, (*d*) appropriate development and exploitation costs to ensure commercial viability and (*e*) approaches for addressing the risk averse nature of the construction industry by ideally entering the construction market through, say, the repair market to provide evidence of performance. It is also important to recognise that the implementation of self-healing construction materials may create a paradigm change in the way that infrastructure is designed and its performance is evaluated. This is also an opportunity as much as a barrier.

5.   The next stage

The promising results from the M4L research on developing self-healing cementitious systems have led to the strategic funding of the much larger research project Resilient Materials for Life (RM4L). While the M4L research reported here was focused specifically on the development of self-healing systems that respond to a physical damage in the form of crack, RM4L, in the next phase of research, has a broader remit focused on tailoring self-healing cementitious systems use in specific commercial applications and on addressing different damage scenarios and conditions. Applications of commercial interest, have been identified through market research (Gardner *et al.*, 2018) and include casting in situ, precast sections, repair, overlays and geotechnical applications. Additionally, important damage scenarios have been identified and include time-related and cyclic damage and chemical damage including corrosion. In addition to self-healing attributes, RM4L will initiate groundbreaking research aimed at embedding self-sensing, self-diagnosing, self-immunisation and self-reporting capabilities into cementitious systems in order to develop truly biomimetic responses in infrastructure materials and structures. Furthermore, RM4L will work closely with industry partners to address challenges associated with the upscaling of the self-healing systems and their introduction to the market. Both numerical modelling and experimental work will be used in the optimisation and tailoring aspects. For more details about the scope and focus of RM4L, see their website (RM4L, 2018).

6.   Conclusions

This paper advocates that those who design and construct the next generation of infrastructure should embrace and use biomimetic cementitious construction materials. This is the mission that the authors and their teams and wider academic and industry partners set themselves when initiating M4L over 4 years ago, and on which they will be engaged for the next 4 years, through RM4L. Through M4L's research, a number of complementary self-healing systems for cementitious matrices have been successfully developed. These include (*a*) microcapsules with switchable mechanical properties and with mineral healing cargo; (*b*) bespoke combinations and microencapsulation of alkaliphilic bacteria, nutrients and precursors for rapid precipitation of healing products; (*c*) innovatively created vascular network using polyurethane tubing; and (*d*) remotely activated SMP multistrand sheathed tendons. Full-scale field trials were undertaken in 2015 to validate the scaling up of the developed systems and their success at delivering self-healing at full scale. There is a significant level of ongoing interest from the industry in taking forward these developments, and associated market research has identified the benefits of self-healing cementitious systems. Cost and economic viability studies provide assurances of whole-life cost-effectiveness of these systems. The work has also identified a number of challenges and barriers, and the next 4 years of research and development will see this national initiative expanded to a completely different level of complexity, breadth and practical application.

Acknowledgements

Financial support from the UK Engineering and Physical Sciences Research Council for the Materials for Life grant (EP/K026631/1, 2013–2016) and the programme grant Resilient Materials for Life (EP/02081X/1, 2017–2022) is gratefully acknowledged. The contribution of the rest of the research team and all the industry partners involved is also gratefully acknowledged.

REFERENCES

Alazhari M, Sharma T, Heath A, Cooper R and Paine K (2018) Application of expanded perlite encapsulated bacteria and growth media for self-healing concrete. *Construction and Building Materials* **160**: 610–619, https://doi.org/10.1016/j.conbuildmat.2017.11.086.

Al-Tabbaa A, Lark R, Paine K, Jefferson T and Embley T (2017) Smart biomimetic construction materials for next generation infrastructure. *Proceedings of ISNGI 2017, London, UK*, pp. 1–10.

Davies RE, Jefferson A and Lark Rand Gardner D (2015) A novel 2D vascular network in cementitious materials. *Proceedings of 2015 fib Symposium, Copenhagen, Denmark*.

Davies R, Pilegis M, Kanellopoulos A *et al.* (2016) Multi-scale cementitious self-healing systems and their application in concrete structures. *Proceedings of the 9th International Concrete Conference 2016 – Environment, Efficiency and Economic Challenges for Concrete, Dundee, UK*.

Davies R, Teall O, Pilegis M *et al.* (2018) Large scale application of self-healing concrete: design, construction and testing. *Frontiers in Materials* **5**: 51, https://doi.org/10.3389/fmats.2018.00051.

De Belie N, Wang J, Bundur ZB and Paine K (2017) Bacteria based concrete. In *Eco-efficient Repair and Rehabilitation of Concrete Infrastructures* (Pacheco-Torgal F, Melchers R, de Belie N *et al.* (eds)). Woodhead, Cambridge, UK, pp. 531–567.

De Belie N, Gruyaert E, Al-Tabbaa A *et al.* (2018) A review of self-healing concrete for damage management of structures. *Advanced Materials Interfaces* **5(17)**: 1800074, https://doi.org/10.1002/admi.201800074.

de Rooij M, Van Tittelboom K, De Belie N and Schlangen E (eds) (2013) *Self-healing Phenomena in Cement-based Materials: State-of-the-art Report of RILEM Technical Committee 221-SHC*. Springer, Dordrecht, the Netherlands.

EC (European Commission) (2014) *Business Innovation Observatory – Smart Living: Advanced Building Materials*. EC, Brussels, Belgium.

Ferrara L, Van Mullem T, Alonso MC *et al.* (2018) Experimental characterization of the self-healing capacity of cement based materials and its effects on the material performance: a state of the art report by COST Action SARCOS WG2. *Construction and Building Materials* **167**: 115–142, https://doi.org/10.1016/j.conbuildmat.2018.01.143.

Gardner D, Jefferson A and Hoffman A (2012) Investigation of capillary flow in discrete cracks in cementitious materials. *Cement and Concrete Research* **42(7)**: 972–981, https://doi.org/10.1016/j.cemconres.2012.03.017.

Gardner D, Jefferson A, Hoffman A and Lark A (2014) Simulation of the capillary flow of an autonomic healing agent in discrete cracks in cementitious materials. *Cement and Concrete Research* **58**: 35–44, https://doi.org/10.1016/j.cemconres.2014.01.005.

Gardner D, Herbert D, Jayaprakash M, Jefferson A and Paul A (2017) Capillary flow characteristics of an autogenic and autonomic healing agent for self-healing concrete. *Journal of Materials in Civil Engineering* **29(11)**: 04017228, https://doi.org/10.1061/%28ASCE%29MT.1943-5533.0002092.

Gardner D, Lark R, Jefferson T and Davies R (2018) A survey on problems encountered in current concrete construction and the potential benefits of self-healing cementitious materials. *Case Studies in Construction Materials* **8**: 238–247, https://doi.org/10.1016/J.CSCM.2018.02.002.

Giannaros P (2017) *Laboratory and Field Investigation of the Performance of Novel Microcapsule-based Self-healing Concrete*. PhD thesis, University of Cambridge, Cambridge, UK.

Giannaros P, Kanellopoulos A and Al-Tabbaa A (2016) Damage recovery in self-healing concrete. *Proceedings of HealCon Conference, Delft, the Netherlands*.

GO-Science (Government Office for Science) (2017) *Technology and Innovation Futures*. GO-Science, London, UK.

HMT (Her Majesty's Treasury) (2016) *National Infrastructure Delivery Plan 2016–2021*. HMT, London, UK. See http://www.gov.uk/government/publications (accessed 31/10/2018).

i3P (Infrastructure Industry Innovation Platform) (2017) *Technology Roadmap for UK Construction & National Infrastructure*. i3P, UK. See https://www.i3p.org.uk/wp-content/uploads/2017/07/i3P-CI-Technology-Roadmap-Booklet-FINAL.pdf (accessed 31/10/2018).

Interesting Engineering (2017) *Simple Self-healing Roads Can Last up to 80 Years*. Interesting Engineering, Istanbul, Turkey. See https://interestingengineering.com/self-healing-roads-last-80-years (accessed 31/10/2018).

Isaacs B, Lark R, Jefferson T, Davies R and Dunn S (2013) Crack healing of cementitious materials using shrinkable polymer tendons. *Structural Concrete* **14(2)**: 138–147, https://doi.org/10.1002/suco.201200013.

Jefferson A, Joseph C, Lark R *et al.* (2010) A new system for crack closure of cementitious materials using shrinkable polymers. *Cement and Concrete Research* **40(5)**: 795–801, https://doi.org/10.1016/j.cemconres.2010.01.004.

Kanellopoulos A, Qureshi TS and Al-Tabbaa A (2015) Glass encapsulated minerals for self-healing in cement based composites. *Construction and Building Materials* **98**: 780–791, https://doi.org/10.1016/j.conbuildmat.2015.08.127.

Kanellopoulos A, Giannaros P and Al-Tabbaa A (2016) The effect of varying volume fraction of microcapsules on fresh, mechanical and self-healing properties of mortars. *Construction and Building Materials* **122**: 577–593, https://doi.org/10.1016/j.conbuildmat.2016.06.119.

Kanellopoulos A, Giannaros P, Palmer D, Kerr A and Al-Tabbaa A (2017) Polymeric microcapsules with switchable mechanical properties for self-healing concrete: synthesis, characterisation and proof of concept. *Smart Materials and Structures* **26(4)**: 045025, https://doi.org/10.1088/1361-665X/aa516c.

Kelly S (2014) The cost of cascading failure: risk and resilience within UK infrastructure networks. *Proceedings of the Future of National Infrastructure Systems & Economic Prosperity Conference, Cambridge, UK.*

Lark RJ, Al-Tabbaa A and Paine K (2013) Biomimetic multi-scale damage immunity for construction materials: M4L Project overview. *Proceedings of 4th International conference on Self-healing Materials, Ghent, Belgium*, pp. 2–5.

Litina C (2015) *Development and Performance of Self-healing Blended Cement Grouts with Microencapsulated mineral Agents.* PhD thesis, University of Cambridge, Cambridge, UK.

ONS (Office for National Statistics) (2016) *Construction Output in Great Britain: May 2016.* ONS, Newport, UK. See https://www.ons.gov.uk/ (accessed 31/10/2018).

Paine K, Lark R and Al-Tabbaa A (2015) Biomimetic multi-scale damage immunity for concrete. Concrete Research: Driving Profit and Sustainability, Jalandhar, India.

Paine K, Alazhari M, Sharma T, Cooper R and Heath A (2016) Design and performance of bacteria-based self-healing concrete. *Proceedings of the 9th International Concrete Conference 2016: Environment, Efficiency and Economic Challenges for Concrete, Dundee, UK*, pp. 545–554.

Pilegis M, Teall O, Hazelwood T *et al.* (2015) Delayed concrete prestressing with shape memory polymer tendons. *Concrete Innovation and Design, fib Symposium, Copenhagen, Denmark.*

Qureshi TS (2016) *The Role of Expansive Minerals in the Autogenous and Autonomic Self-healing of Cement Based Materials.* PhD thesis, University of Cambridge, Cambridge, UK.

RM4L (Resilient Materials 4 Life) (2018) http://rm4l.com/about/ (accessed 31/10/2018).

Sharma TK, Alazhari M, Heath A, Paine K and Cooper RM (2017) Alkaliphilic *Bacillus* species show potential application in concrete crack repair by virtue of rapid spore production and germination then extracellular calcite formation. *Journal of Applied Microbiology* **122(5)**: 1233–1244, https://doi.org/10.1111/jam.13421.

Souza LR (2017) *Polymeric Microcapsules Using Microfluidics for Self-healing in Construction Materials.* PhD thesis, University of Cambridge, Cambridge, UK.

Teall O (2017) *Crack Closure and Enhanced Autogenous Healing of Structural Concrete Using Shape Memory Polymers.* PhD Thesis, Cardiff University, UK.

Teall O, Davies R, Pilegis M *et al.* (2016) Self-healing concrete full-scale site trials. In *Proceedings of the 11th fib International PhD Symposium in Civil Engineering, FIB 2016* (Maekawa K, Kasuga A and Yamazaki J (eds)). Balkema, Rotterdam, the Netherlands, pp. 639–646.

Teall O, Pilegis M and Davies R (2017) Development of high shrinkage polyethylene terephthalate (Pet) shape memory polymer tendons for concrete crack closure. *Smart Materials and Structures* **26(4)**: 045006, https://doi.org/10.1088/1361-665X/aa5d66.

Teall O, Pilegis M, Davies R *et al.* (2018) A shape memory polymer concrete crack closure system activated by electrical current. *Smart Materials and Structures* **27(7)**: 075016, https://doi.org/10.1088/1361-665X/aac28a.

Tilly G and Jacobs J (2007) *Concrete Repairs: Performance in Service and Current Practice.* BRE Press, Bracknell, UK.

WEF (World Economic Forum) and BCG (The Boston Consulting Group) (2016) *Shaping the Future of Construction: a Breakthrough in Mindset and Technology.* WEF, Cologny, Switzerland.

Dhir and Paine
ISBN 978-0-7277-6457-7
https://doi.org/10.1680/icetsc.64577.103
ICE Publishing: All rights reserved

Chapter 7

Durability of self-healing ultra-high-strength reinforced micro-concrete under freeze–thaw or chloride attack

Gloria Pérez
Researcher, Eduardo Torroja Institute for Construction Sciences – CSIC, Madrid, Spain
(corresponding author: gperezaq@ietcc.csic.es)

José Luis García Calvo
Researcher, Eduardo Torroja Institute for Construction Sciences – CSIC, Madrid, Spain

Pedro Carballosa
Researcher, Eduardo Torroja Institute for Construction Sciences – CSIC, Madrid, Spain

Run Tian
Predoctoral fellow, School of Engineering, Hong Kong University of Science and
Technology, Hong Kong

Virginia Rodriguez Allegro
Researcher, Eduardo Torroja Institute for Construction Sciences – CSIC, Madrid, Spain

Edurne Erkizia
Senior researcher, Tecnalia Materials, Sustainable Construction Division, Derio, Spain;
MATCON, Associated Unit CSIC-Tecnalia, Madrid, Spain

Juan Jose Gaitero
Senior researcher, Tecnalia Materials, Sustainable Construction Division, Derio, Spain;
MATCON, Associated Unit CSIC-Tecnalia, Madrid, Spain

Ana Guerrero
Senior researcher, Eduardo Torroja Institute for Construction Sciences – CSIC, Madrid,
Spain

The durability of an innovative self-healing micro-concrete in aggressive environments simulated by freeze–thaw cycles and salt spray test is studied. Tests are conducted on an ultra-high-strength reinforced micro-concrete that incorporates an autonomous self-healing mechanism based on the reaction of an epoxy compound enclosed within silica microcapsules and amine functionalised silica nanoparticles distributed within the cementitious matrix. The effect of aggressive environments is analysed in the self-healing micro-concrete and in a reference micro-concrete stored for 28 d in laboratory conditions after cracking, for crack widths of 150 and 300 μm. The results of capillary water absorption tests, complemented by electron microscopy analysis, confirm the enhanced durability of the autonomously self-healed material in both the freeze–thaw and the salt spray tests, as compared to the reference micro-concrete. In conclusion, the innovative self-healing mechanism is expected to increase the service life of structures in humid, cold climates and chloride-containing environments.

Introduction

Concrete is one of the most widely used construction materials in the world because of its relatively low cost and high compressive strength. However, its limited tensile strength and brittle nature make it prone to the formation of cracks that reduce the durability of the material, causing damage and even collapse of concrete structures. Although crack repair is possible, it usually implies a high cost and it becomes difficult when cracks cannot be seen or are unreachable. Cracks also provide preferential routes for the ingress of aggressive agents into the concrete. For example, in marine environments the penetration of chloride into the concrete matrix is accelerated, thus limiting the durability of reinforced concrete structures. Taking this into account, autonomous self-healing cement-based materials have been developed in recent times, and different strategies have been considered (Huang *et al.*, 2016; Van Tittelboom and De Belie, 2013). One of these innovative self-healing systems, based on the addition of a two-component epoxy-amine adhesive, has been developed and introduced to the cementitious matrix as described in previous works (Pérez *et al.*, 2015a, 2015b). One of the components, the epoxy compound, is inserted encapsulated in silica microcapsules, whereas the other component, the amine group, is introduced by adding amine functionalised silica nanoparticles, which react with the cementitious matrix and thus incorporate the amine group within it. Hence, when a microcrack occurs and propagates, it breaks some of the microcapsules and the epoxy contained within is released to the crack, coming into contact with amine groups dispersed within the cement matrix. Upon reaction of the two components of the adhesive, the epoxy hardens within the crack and provides a sealing effect. The compatibility and stability of these additions in the cementitious matrix has been studied in a previous work (Pérez *et al.*, 2015b).

The efficiency of the proposed self-healing system has been confirmed in an ultra-high-performance reinforced micro-concrete (UHPRC) designed with a very low water/cement ratio, a high cement content, silica fume and steel fibres (García Calvo *et al.*, 2017). Materials of this type are expected to comply with especially demanding performance and durability requirements, thus implementation of an autonomous self-healing system to enhance the inherent autogenous crack-sealing capacity of UHPRC is proposed to obtain the healing of wider cracks, preventing their growth and limiting the ingress of aggressive agents. UHPRC with such an enhanced self-healing capacity is especially valuable to assure durable concrete structures under severe environmental and operating conditions, such as in underground applications (high pressures with high-temperature gradients) and in marine environments (with high chloride concentrations) or arctic areas (with low temperatures and the action of ice). In these highly demanding scenarios, the initial cost increase associated with the implementation of a self-healing system is widely compensated within the context of a life-cycle analysis by the reduction of maintenance cost and extension of service life that are obtained.

Taking this into account, the present work focuses on the efficiency of the proposed self-healing system in UHPRC under aggressive environments. Within this context, one of the main causes of concrete deterioration is the freeze–thaw action, especially in cold climates with high humidity and low temperatures; this affects concrete specimens' tightness and consequently their permeability to aggressive agents (Bogas *et al.*, 2016; Gonzalez *et al.*, 2016; Medina *et al.*, 2013; Pentalla, 2006; Richardson *et al.*, 2016; Sabir, 1997; Tikkanen *et al.*, 2015; Vegas *et al.*, 2009).

According to Pentalla (2006), when wet concrete is submitted to the first freezing step, the pore water transfers towards the newly formed ice lenses to compensate for the different chemical potential of the two phases and the sample shrinks. Upon cooling down to around −20°C, water freezes even in the small capillary pores of concrete and ice is formed into the pore space to an extent that prohibits expansion of the newly forming ice. In this situation, a 9% volume expansion is associated with ice formation; if the hydraulic pressure cannot be relieved, the water pressure will expand the pores, causing tensile stresses in the surrounding concrete paste. In saturated concrete, the tensile stress may eventually exceed the tensile capacity of the paste and cracking will occur (Hale *et al.*, 2009).

Mass loss and surface scaling of the specimens are usually analysed to define the external damage caused by the cycles; some of the most relevant variables influencing this scaling in the case of UHPRC are the water/cement ratio of the matrix and the curing time of the specimens (Pentalla, 2006). However, ultrasonic pulse velocity and mechanical strength variations are measured for the analysis of internal damage. Although tests investigating the resistance of UHPC to freeze–thaw have been limited, the related studies have demonstrated their high durability in this type of environment (Acker and Behloul, 2004; Ahlborn *et al.*, 2008; Alkaysi *et al.*, 2016).

Recent research works related to freeze–thaw resistance of concrete are mainly concerned with the effect of different constituents, often secondary materials, as additions to cement (Richardson *et al.*, 2016), as fine fillers (Tikkanen *et al.*, 2015) or as aggregates (Bogas *et al.*, 2016; Medina *et al.*, 2013) for development of more sustainable cement-based materials with high durability. Also newly developed techniques are proposed to give a new insight into the variation of porosity and pore volume upon freeze–thaw cycles (Yuan *et al.*, 2016) or for the in situ determination of internal degradation of structures submitted to cold, humid climates (Lu *et al.*, 2015). Regarding self-healing concretes, Zhu *et al.* (2012) analysed the efficiency of autogenous self-healing in a cementitious material when sealing of cracks occurs under freeze–thaw cycles, but, to the current authors' knowledge, the present work is the first to study the durability of a sealed cementitious material with an autonomous self-healing system in this aggressive environment.

In order to perform a complete analysis of the efficiency of the proposed self-healing micro-concrete under aggressive environments, the effect of chloride ingress must be taken into account, as cracks often act as preferential paths, giving rise to higher chloride attack upon the appearance of cracks (Berrocal *et al.*, 2015; Kim *et al.*, 2014). In fact, this is known to be one of the major durability issues in concrete infrastructures, especially in marine environments, where access to constructions is usually difficult. Consequently, fast crack sealing without human intervention through the addition of an autonomous self-healing system within the concrete is especially valuable in this case.

Several test methods have been proposed to study chloride ingress in concrete, including immersion tests, migration tests or resistivity tests (Andrade *et al.*, 2013). Related to this subject, much recent research may be found in the literature related to the effect of different parameters on rebar corrosion in reinforced concrete by chloride ingress, such as the water/cement ratio, the presence of cracks and the crack width, or the addition of fibres and pozzolanic additives (Berrocal *et al.*, 2015; Kim *et al.*, 2014; Kwon *et al.*, 2009). Some authors also work on the durability of recently developed concrete materials, such as self compacting concrete (SCC) (Ryan and O'Connor, 2016), and on modelling chloride concentration and ingress in order to obtain service life prediction of concrete in chloride-bearing environments (Andrade *et al.*, 2013; Kwon *et al.*, 2009; Seify *et al.*, 2016). Nevertheless, as in the case of freeze–thaw resistance, studies on the durability of self-healing concrete in chloride environments are very scarce. Maes *et al.* (2014) studied autonomous crack healing of mortar specimens with embedded polyurethane capsules and found that the results are influenced by the particular behaviour of the self-healing system in each specimen; also, when the healing mechanism works properly and polyurethane seals the crack efficiently, no chlorides penetrate along the crack. This indicates that an important service life extension of concrete structures in chloride environments can be obtained when an efficient self-healing system is incorporated into the concrete matrix.

In the case of the self-healing system studied in the present work, preliminary results on its efficiency when cracked specimens are subjected to aggressive environments immediately after cracking have been presented previously (Guerrero *et al.*, 2015; Pérez *et al.*, 2015c). However, these preliminary results did not consider the healing time in the evaluated efficiency. Thus, the present work studies the performance under freeze–thaw cycles (FT) and under chloride ingress during salt spray tests (SS) of the proposed innovative self-healing micro-concrete once healed under controlled laboratory conditions for 28 d after cracking. The results of this analysis are important to prove whether the self-healing system proposed improves the durability and increases the service life of the cementitious material in cold climates, with high humidity and low temperatures, and in marine environments with high chloride attack.

Experimental set-up

Specimens' preparation and cracking

Cylindrical specimens, 100 mm in diameter and 25 mm thick, of ultra-high-performance micro-concretes were prepared with a water/cement ratio of 0·3. The mix proportions, inspired by the work of Richard and Cheyrezy (1995), are shown in Table 1 for both self-healing (SH) and reference (Ref) micro-concrete. The contents of self-healing additions in the SH micro-concrete were chosen according to the good self-healing efficacy and good mechanical performance shown by this micro-concrete mix in previous works (García Calvo *et al.*, 2017; Pérez *et al.*, 2015b).

To prepare the specimens, the silica fume Elkem Grade 940-U undensified (SF), microcapsules encapsulating the epoxy (CAP), aggregates, limestone powder and cement were mixed slowly (according to UNE-EN 196-1 (UNE, 2005)) for 1 min. The amine-functionalised nanosilica (NS) had been previously dispersed with an ultrasonic probe in distilled water with the superplasticiser Structuro 351 from Fosroc; the resulting mixture was then added to the solids. All the materials were mixed for 1 min at a slow rate, 3 min at a fast rate, then halted for 2 min, followed by a further 2 min at the fast rate. Then, half of the steel fibres (OL13/0·16) were added and mixed

Table 1 Mix proportions for specimens

	Reference (Ref)	Self-healing (SH)
CEM I 52·5N: kg/m^3	885	885
Water: kg/m^3	266	266
Siliceous aggregates $\varnothing < 1$ mm: kg/m^3	664	664
Limestone aggregates $\varnothing < 1$ mm: kg/m^3	664	664
Superplasticiser: kg/m^3	35·4	35·4
Steel fibre: kg/m^3	155	155
Silica fume, SF: kg/m^3	221	119
Epoxy encapsulating silica microcapsules, CAP: kg/m^3	0	44·3
Amine functionalised nanosilica, NS: kg/m^3	0	58·6

rapidly for 1 min, followed by the addition of the remaining steel fibres and fast mixing for another minute. The specimens were stored in a controlled chamber at $20 \pm 1°C$ and 100% relative humidity and were demoulded after 24 h, then cured under water at $20 \pm 1°C$.

At the age of 28 d, a set of three samples of the Ref micro-concrete and three samples of the SH micro-concrete were subjected to a capillary water absorption test in order to provide a reference of the tightness of the specimens before cracking. The remaining specimens were cracked using a splitting tension test, controlling and measuring the width of crack mouth opening displacement (CMOD) by an extensometer placed between two steel plates stuck on one face of the specimen. The tests were conducted at a constant rate of crack mouth opening of $0·5$ μm/s and two final CMODs were used: 150 μm and 300 μm. When promoting the defined CMODs, crack relaxation took place upon unloading in all cases and the final crack width (i.e. the remaining CMOD) changed from Ref samples to SH samples. In particular, for the case of 150 μm CMOD, the remaining crack width had a mean value of 60 μm in Ref samples and 90 μm in SH samples. For the case of 300 μm CMOD, the remaining crack width had a mean value of 190 μm in Ref samples and 235 μm in SH samples.

The cracked specimens were divided into three sets, as indicated in the scheme of Figure 1. Each set was formed by six samples of Ref micro-concrete, three cracked up to a CMOD of 150 μm and three cracked up to a CMOD of 300 μm, and six samples of SH micro-concrete with the same distribution. All the specimens were stored in sealed plastic bags at 20°C after cracking to prevent carbonation and to prohibit moisture exchange with the environment during the healing period, so that any self-healing that occurred was due to the epoxy-amine self-healing system acting under ideal conditions. After a healing time of 28 d in such conditions, one set of Ref and SH specimens (termed LC) was subjected to a capillary water absorption test. Another set (termed FT) was subjected to the freeze–thaw test and the third set of specimens (termed SS) was exposed to the salt spray test, in order to evaluate the durability of the healed material. These two sets of specimens were subjected to the capillary water absorption test immediately after the corresponding durability test.

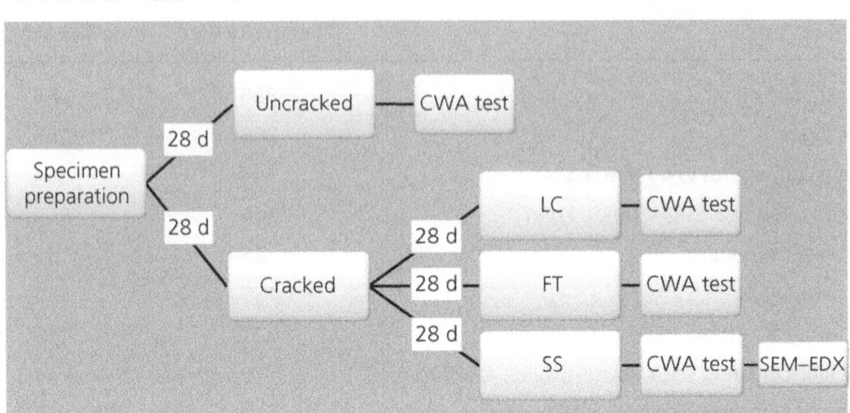

Testing procedures

The specimens of the FT set were weighed before placing them on a metal support grating inside a climate chamber to be subjected to a freeze and thaw durability test consisting of cycles of 3 h in dry ambient conditions at −20°C and 3 h under water at 20°C. A similar temperature cycle has been used by other authors (Pentalla, 2006; Vegas *et al.*, 2009) in the analysis of concrete durability under freeze–thaw action and, in fact, this complies with the conditions stated in ASTM C666 standard test method (procedure B) (ASTM, 2003), taking into account that the temperature at the core of the specimens must be lower than the experimental chamber temperature during thawing phases (Vegas *et al.*, 2009). The freeze–thaw test was carried out until a total number of 154 cycles had been reached and, in order to determine external damage through scaling, the aspect of the specimens' surface was controlled and their weight was measured several times during the test.

The specimens of the SS set were weighed before placing them in a climatic chamber that complied with the UNE-EN-ISO9227-2007 (UNE, 2007) standard to be subjected to the salt spray test with a sodium chloride (NaCl) solution of 5% concentration at 23°C for a period of 82 d. The chamber was periodically opened, after having been previously filled with compressed air, in order to weigh and observe the samples to record external variations. After this inspection, the samples were placed into the chamber again, changing their position to homogenise the results, in order to proceed with the test.

According to Ryan and O'Connor (2016), this kind of salt spray test represents a form of laboratory testing that is closer to real marine environments and may be considered as more suitable for different building materials than other methods usually reported in the literature that involve an electrical acceleration of the chloride ingress.

A capillary water absorption test is performed to analyse the internal damage produced by the aggressive environments in the micro-concrete specimens. This test is usually done to study the recovery of tightness due to self-healing strategies in cement-based materials (Muhammad *et al.*, 2016; Tang *et al.*, 2015; Van Tittelboom and De Belie, 2013) and has also proven to be useful in quantifying the degree of recovery in preloaded specimens exposed to freeze–thaw cycles for autogenous self-healing (Zhu *et al.*, 2012).

In this work, the samples were dried at 50°C until constant weight as a preconditioning procedure for the measurement of water absorption rate and capillary suction coefficient by means of a capillarity test based on the standard UNE 83982 (UNE, 2008). According to Yang (2016), this drying temperature would produce an increase of accessible porosity in concrete as compared to drying at lower temperatures, but avoids the microstructural alterations produced by higher temperatures. Once dried, the specimens were placed in a receptacle with a water level of 5 mm, with only one face in contact with the water. Test specimens were periodically weighed and their mass increase was measured in order to determine the water absorption rate and the capillary absorption coefficient.

A scanning electron microscope (SEM) Hitachi S-4800 equipped with an energy dispersive X-ray (EDX) analyser (Bruker 5030) was used for morphological and chemical analysis of certain samples (see Figure 1). The samples were embedded in an epoxy resin and polished. The resulting surface was metallised with gold before the analysis in back-scattering mode.

Results and discussion

Freeze–thaw test

The weights of the specimens of Ref and SH micro-concretes were measured after certain cycles, always during the thawing phase, in order to assure a similar degree of saturation in all the measurements. The resulting mean values of the percentage weight change from the initial value for samples of the FT set are depicted in Figure 2, together with the standard deviation.

There is a significant weight increase of specimens during the initial cycles, which can be associated with the water saturation during the underwater step at 20°C of the freeze–thaw cycles. A significant variation is observed in this initial weight gain for the different types of specimens in the FT set, with a higher weight increase in the samples of reference micro-concrete cracked up to 300 μm, which show a 0·50% mean increase at the first weight measurement, after the first 58 cycles, whereas similar samples of self-healing micro-concrete

Figure 2 Percentage weight change of the specimens of FT set with respect to their initial weight during the freeze–thaw durability test

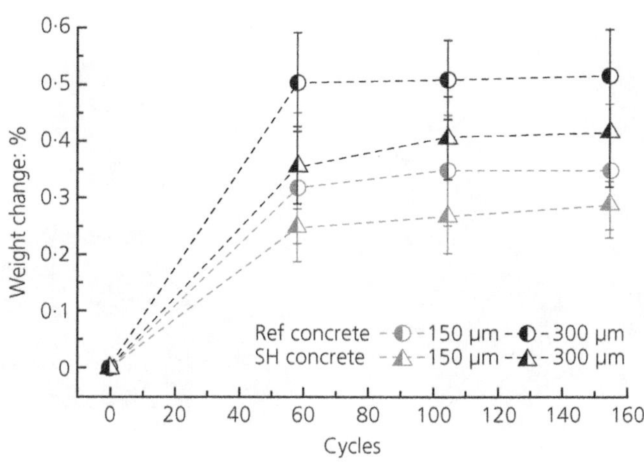

show a mean increase of 0·35%. A similar trend is also observed for specimens cracked up to 150 μm, indicating a higher amount of water entering the reference specimens than is the case for self-healing micro-concrete specimens with the same nominal crack width. These results must be related to the expected sealing of cracks during the 28 d storage prior to the test in the material that incorporates the self-healing system.

After the initial cycles and up to 154 cycles, the variation of weight change is lower than 0·1% for all samples, indicating a constant weight of the specimens. Moreover, the physical aspect of the samples was also controlled during the test and no significant external damage was observed (see Figure 3). Thus, it may be concluded that both the reference and self-healing micro-concretes analysed achieve saturation and show no scaling due to the cycles.

Figure 3 Surface aspect of specimens with CMOD of 300 μm prior to the freeze–thaw test: (a) SH concrete; (b) Ref concrete. Surface aspect of the same specimens at the end of the 154 cycles: (c) SH concrete; (d) Ref concrete

(a)

(b)

(c)

(d)

In order to evaluate the efficacy of the self-healing mechanism under the freeze–thaw cycle conditions, a capillary water absorption test was undertaken after 154 cycles. The mass increase data during the water absorption test for reference specimens stored in sealed plastic bags at standard laboratory conditions for 28 d (Ref-LC) are compared in Figure 4(a) with the data for the reference samples subjected to the freeze–thaw cycles after the same storage conditions (Ref-FT). A similar comparison is presented in Figure 4(b) for the case of self-healing micro-concrete (SH-LC and SH-FT) specimens, with the data for corresponding uncracked samples also included in each figure for comparison. Each data point is the mean

Figure 4 Mean mass increase during water absorption test of (a) Ref concrete and (b) SH concrete subjected to FT test. In both cases, the results for similar samples stored in laboratory conditions (LC) and for uncracked samples are included for comparison

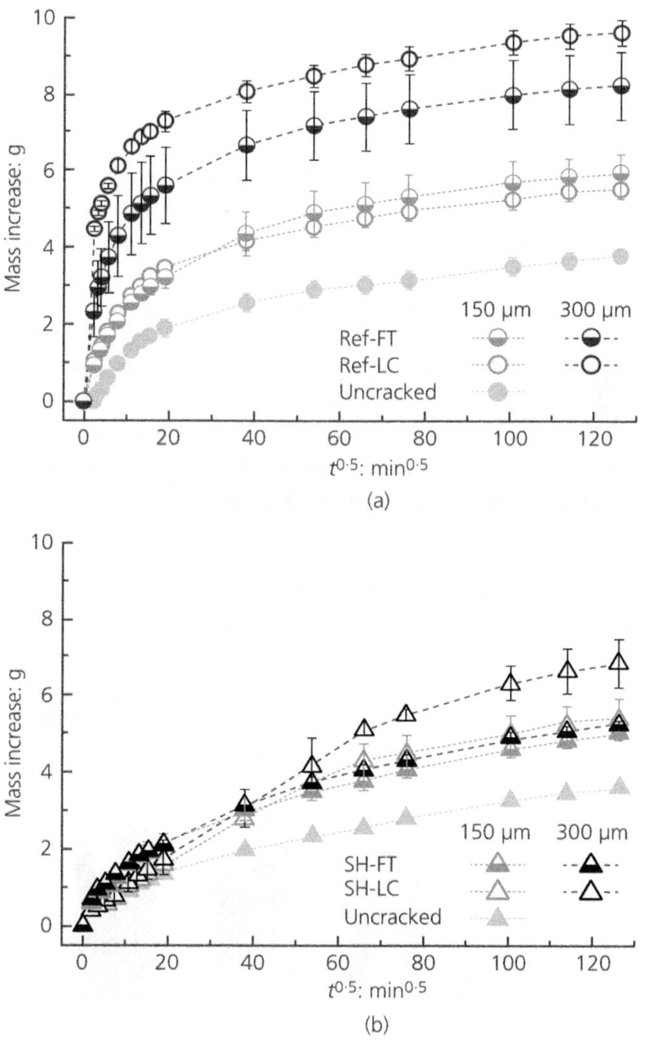

value of three equivalent specimens and the error bars represent the standard deviation of the data. Figure 5 shows the corresponding capillary water absorption coefficients after 24 h of testing.

For the reference micro-concrete, the initial water absorption rate of the cracked specimens is, as expected, significantly higher than that observed in uncracked samples, and increases with the increase of the nominal crack width from 150 to 300 μm. From the behaviour of the reference samples with CMOD of 150 μm under the capillary absorption test (Figure 4(a)), no significant influence of freeze–thaw cycles may be inferred during the initial test period for the FT set. In fact, the mean capillary absorption coefficient K (see Figure 5) of samples Ref-LC after 24 h (time$^{0.5}$ equal to 38 min$^{0.5}$) is 0·017 kg/m^{2}min$^{0.5}$, clearly higher than the value for the uncracked reference micro-concrete (0·008 kg/m^{2}min$^{0.5}$), but very similar to the value of 0·015 kg/m^{2}min$^{0.5}$ obtained for Ref-FT.

However, in the case of the specimens with the CMOD of 300 μm, the water absorption rate of Ref-FT specimens, with K value of 0·022 kg/m^2, is lower than the Ref-LC specimens, with K value of 0·027 kg/m^2. Furthermore, a significantly lower final absorbed water amount is observed for Ref-FT in Figure 4(a), which may be associated with permeation in the underwater thawing period promoting secondary hydration within the cement matrix that benefits from autogenous healing. This effect must be enhanced for the higher CMOD value and compensates for any reduction of tightness that could be induced by the freeze–thaw cycles.

Regarding the self-healing micro-concrete specimens stored in laboratory conditions (SH-LC in Figure 4(b)), a lower mass increase with respect to the reference micro-concrete (Ref-LC in Figure 4(a)) is observed, especially in the case of CMOD 300 μm. In fact, the initial slope of the mass increase curve for SH-LC cracked samples is very similar to that obtained for uncracked samples for both crack widths. This result clearly confirms that healing is effective in samples maintained for 28 d in laboratory conditions, although this healing is not complete, as suggested by the higher final mass increase observed in cracked samples (García Calvo *et al.*, 2017).

Figure 5 Capillary water absorption coefficient of Ref and SH concrete sets stored in laboratory conditions (LC) and subjected to durability tests (see text for notation)

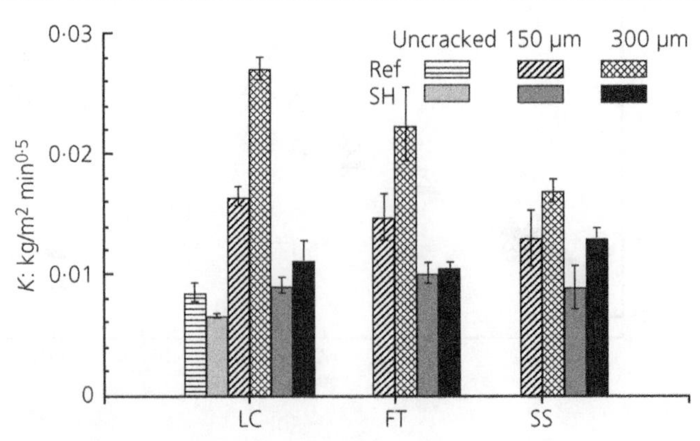

The results for the freeze–thaw cycles conducted on self-healing micro-concrete specimens 28 d after cracking (SH-FT) show a very similar rate in the initial mass increase to the rate for uncracked specimens. After 24 h (time$^{0.5}$ equal to 38 min$^{0.5}$), the mass increase of SH-LC and SH-FT sets at both crack widths takes a comparable value, slightly higher than the value corresponding to the uncracked specimens. This behaviour confirms that tightness recovery by self-healing is not degraded under freeze–thaw attack. For longer test times, the only effect observed is a lower mass increase in the specimens of the SH-FT set cracked to 300 μm as compared to similar samples stored in laboratory conditions. The results indicate that a densification of the SH-FT set with respect to the SH-LC set associated with secondary hydration in the thawing cycles is also present in this case.

In summary, the results suggest that the immersion in water during the thawing cycles gives rise to an enhancement of the autogenous healing, both in reference and self-healing micro-concretes, which is highly effective due to the low water/cement ratio of the blends. This conclusion is in agreement with the enhancement of the autonomous healing by an autogenous mechanism in bacteria-based self-healing concrete due to immersion in water during wet–dry cycles reported by Tziviloglou *et al.* (2016).

More interesting is to highlight that not only is the durability of the micro-concrete not reduced by the use of the healing additions, but the sealing obtained by the self-healing system is effective under the extreme conditions imposed during the freeze–thaw cycles. In fact, the capillary absorption coefficient K of self-healing material after freeze–thaw cycles (Figure 5) takes a comparable value for the case of specimens cracked to CMOD of both 150 μm and 300 μm (0·010 and 0·011 kg/m^{2}min$^{0.5}$). These values are 33% and 50% lower than those obtained for the reference material specimens with 150 μm and 300 μm crack width, respectively. This higher tightness of self-healing micro-concrete submitted to freeze–thaw attack assures an improvement in the material durability under this aggressive environment.

Salt spray test

During the salt spray test, the specimens were periodically weighed and inspected in order to detect possible damage caused by the severe exposure conditions. However, no significant scaling was observed on the matrix of the specimens during the test. A darker tonality was observed in the central surface of reference micro-concrete specimens, which may be related to a higher saline solution penetration in these specimens, thus corroborating a slightly higher water absorption coefficient of Ref samples. Regarding the steel fibres of the specimens' surfaces, they were oxidised in all cases after the test period (see Figure 6).

Figure 7 shows the mean values of the percentage weight change measured during the salt spray test for the SS set. There is a significant weight increase of specimens during the initial days of the test that may be attributed to penetration of the pulverised saline solution and also to the formation of corrosion products in the fibres. The samples of reference micro-concrete show a mean weight increase during the first 7 d of 1·4 and 2·1% for crack widths of 150 and 300 μm, respectively. For the same test period, the corresponding samples of self-healing micro-concrete subjected to SS test show a 1·0 and 1·6% mean increase. These results confirm the improvement of tightness due to the sealing of cracks during the 28 d storage of specimens

Figure 6 Surface aspect of (a) Ref and (b) SH concrete specimens after the SS test (CMOD of 300 μm)

(a)   (b)

Figure 7 Percentage weight change of the specimens of SS set with respect to their initial weight during the salt spray test

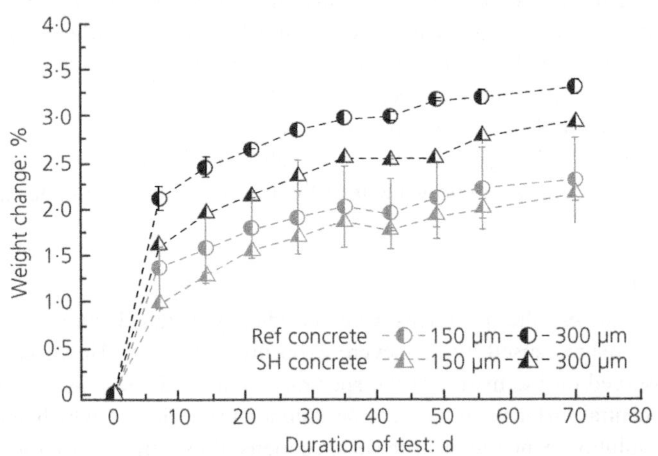

prior to the salt spray test in the self-healing cementitious material. As the test proceeds, a clear increase of the measured weight with time is observed that may be due to further reaction between saline solution and samples.

The mean mass increase values of the specimens subjected to the subsequent capillary absorption test are calculated to analyse the effect of the salt spray test on the self-healing system. The data for specimens subjected to the SS test are compared with the respective samples stored in sealed plastic bags at standard laboratory conditions (LC) in Figure 8 and the data for uncracked samples are included for comparison. The mass increase of Ref-SS samples

Figure 8 Mean mass increase during water absorption test of (a) Ref concrete and (b) SH concrete submitted to SS test. In both cases, the results for similar samples stored in laboratory conditions (LC) and for uncracked samples are included for comparison

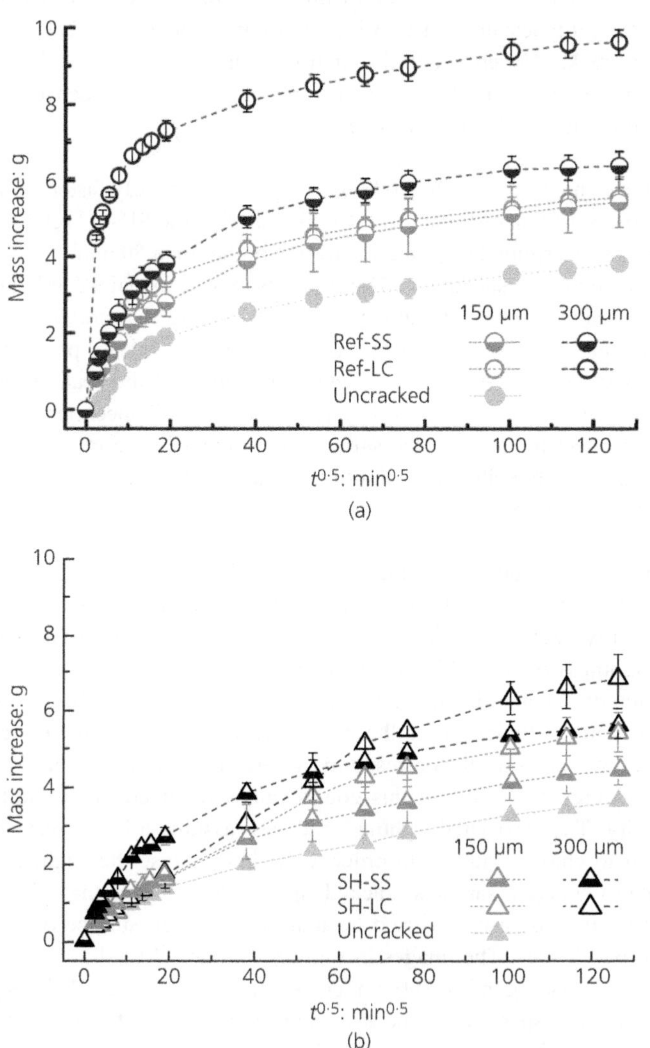

with CMOD of 300 μm is clearly lower than that of corresponding Ref-LC samples (Figure 8(a)). In fact, the capillary absorption coefficient (K) after 24 h of samples with this crack width (shown in Figure 5) is 0·017 and 0·027 kg/m^2, for Ref-SS and Ref-LC, respectively. The same tendency is observed for the corresponding samples with the CMOD of 150 μm in the initial part of the test, but with a less significant difference. This suggests that, as well as in freeze–thaw cycles, the salt spray test promotes an increased tightness of the cracked specimens, related in this case to water permeation promoting the secondary hydration of the micro-concrete and/or precipitation of reaction products due to the saline environment.

Self-healing micro-concrete specimens cracked to a CMOD of 150 µm and stored for 28 d before the salt spray test (SH-SS) show a similar mass increase rate during the initial stages of the water absorption test as the corresponding SH-LC set (Figure 8(b)). In fact, the value of the absorption coefficient at 24 h is 0·009 kg/m^{2}min$^{0·5}$ for both sets of cracked specimens, indicating that the durability test under a chloride environment does not affect the effective sealing of these thinner cracks. Nevertheless, the final mass increase is clearly lower in the case of SH-SS specimens compared to the SH-LC set, which confirms the densification induced by autogenous healing during the durability test.

In the case of the samples cracked to a CMOD of 300 µm, a clearly faster initial mass increase is obtained for the SH-SS set of specimens as compared to the SH-LC set. On the contrary, as the capillary water absorption test proceeds, for $t^{0·5}$ higher than 50 min$^{0·5}$, the mass increase is clearly higher in the SH-LC specimens. These results may be explained as being attributable to a non-complete sealing of the 300 µm wide cracks during the storage time in laboratory conditions. The saline solution may penetrate these cracks that are partially filled with the epoxy-amine adhesive, giving rise to a slight degradation of the sealing that results in the fastest initial mass increase observed in Figure 8(b) for SH-SS specimens. At the same time, the densification effect induced by the saline environment in the micro-concrete matrix, as observed in samples with well-healed, thinner cracks, must be responsible for the lower final mass increase in the SH-SS set.

The results indicate the stability of the material healed 28 d after cracking under the salt spray test conditions for a CMOD of 150 µm, whereas the tightness of the healed material seems to be slightly degraded in the saline environment in the case of a CMOD of 300 µm. Even so, the tightness of the self-healing material is, also in this case, clearly higher than that of the reference material. For self-healing specimens with 150 µm crack width, the capillary absorption coefficient after the salt spray test depicted in Figure 5 is 0·009 kg/m^{2}min$^{0·5}$, 31% lower than for the corresponding reference specimens. In the case of 300 µm crack width, a decrease of 24% in this coefficient is obtained upon addition of the self-healing capability. The associated tightness increase assures a better resistance of the self-healing material to chloride attack. In order to further analyse the slight effect observed in self-healing micro-concrete samples cracked up to 300 µm after the salt spray test, back-scattered electron microscopy was used to analyse the crack surface obtained by splitting specimens into two halves. The images in Figures 9(a) and 9(b) show the general aspect of the surface after polishing of a 300 µm crack, at approximately half depth of samples REF-SS and SH-SS, respectively. The expected morphology of the reinforced micro-concrete matrix is observed in both cases, with the additional circular structures corresponding to the capsules at positions 1 and 2 in the image of the self-healing material. The capsules may be distinguished from round, hollow pores by the presence of a dark material filling the capsules that corresponds to the epoxy compound and also for the clear circular boundary around the silica capsules corresponding to the bonding with the surrounding cementitious matrix (Pérez *et al.*, 2015b). Nevertheless, the most interesting difference between the two samples is observed on the circular or oval-shaped elements corresponding to the fibres. The expected uniform surface is shown in the case of the SH-SS sample, whereas crystalline structures are clearly observed to have developed on the fibres in the Ref-SS sample in Figure 9(a) and in the zoomed images of the fibre marked by a star in Figure 9(a), as shown

Figure 9 Backscattering electron microscopy images of the polished crack surface of samples cracked to 300 µm of (a) Ref-SS and (b) SH-SS sets. Zoomed images corresponding to the area marked by a star in (a) are included in images (c) and (d)

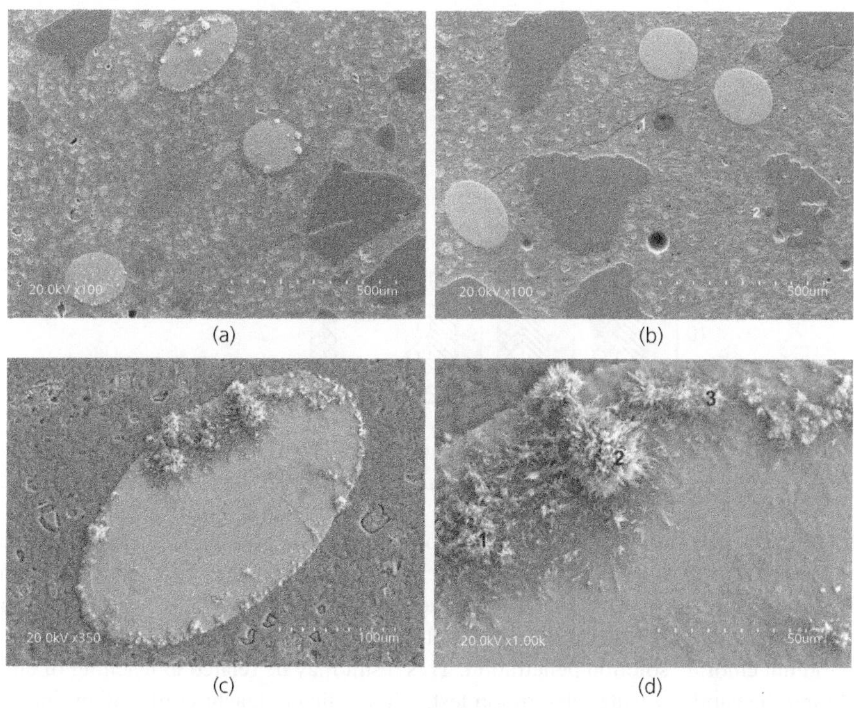

(a)   (b)

(c)   (d)

in Figures 9(c) and 9(d). Significant differences are observed in the composition obtained by EDX analysis for different points in the crystalline structures observed in Figure 9(d). Taking into account that similar morphologies with similar compositional variations are encountered in different fibres throughout the sample surface, individual EDX spectra were measured in at least ten points within each of the areas marked as 1, 2 and 3 in Figure 9(d) in order to obtain representative compositional results. The resulting normalised mean atomic concentrations are collected in Figure 10, excluding the contribution from the gold (Au) conductive coating used in the sample preparation and representing only elements with a percentage contribution higher than 1%.

As seen in Figure 10, the crystals of areas 1 and 2 have similar compositions, with a percentage atomic concentration of oxygen atoms around three times the concentration of carbon, which is similar to that of sodium. This suggests that the crystals correspond to sodium carbonate phase incorporating also potassium atoms, probably coming from the concrete matrix, and iron atoms coming from the fibre composition. In the case of area 3 in Figure 9(d), the composition obtained by EDX is also coherent with a carbonate phase, but in this case incorporating a higher contribution of potassium (K) as compared to sodium (Na) and also a copper (Cu) atomic contribution higher than 1% coming from the fibres.

Figure 10 Atomic percentage composition of the areas marked as 1, 2 and 3 in Figure 9(d), as obtained by EDX analysis (C, carbon; O, oxygen; Na, sodium; K, potassium; Fe, iron; Cu, copper)

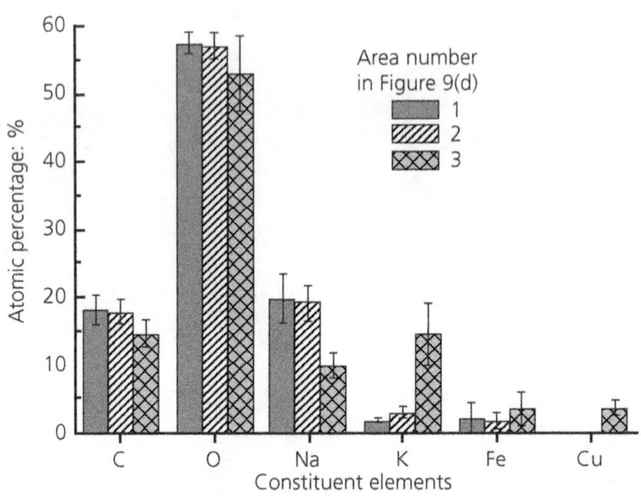

The high atomic percentage of sodium atoms observed in the crystals indicates that the saline solution employed to simulate the aggressive marine environment has entered the sample matrix through the crack. Also interesting is to highlight that chloride is not detected in the crack surface of the Ref-SS sample, as seen in the representative image of Figure 9(d), in spite of the sodium chloride solution penetration. This result may be related to leaching of chloride ions during the capillary water absorption test, while efficient leaching of sodium ions is not observed owing to their incorporation into the crystalline carbonate phase.

Finally, the durable sealing of the crack in the case of self-healing micro-concrete that prevents the entrance of the saline solution is confirmed by the analysis, as sodium-containing crystals are not present in the SH-SS sample analysed (Figure 9(b)). These results are coherent with those reported by Maes *et al.* (2014) indicating that, when an autonomous self-healing mechanism works properly, no chlorides penetrate along the sealed crack.

In conclusion, according to the results obtained, it is confirmed that the autonomous self-healing mechanism tested in this work provides an increase in tightness of the specimens submitted to the salt spray test. This gives rise to an enhanced resistance to sodium chloride penetration into the material, which is expected to increase the service life of marine reinforced concrete infrastructures built with the proposed self-healing micro-concrete.

Conclusions

The durability of an ultra-high-strength micro-concrete with an innovative autonomous self-healing mechanism based on an epoxy-amine adhesive is studied in this work. As mentioned in the introduction, the efficient sealing of cracks with a width up to 300 μm had already been proven in samples of this self-healing micro-concrete stored in standard environmental conditions during 28 d after crack promotion. In the present study, the effect of aggressive environments (freeze and thaw action and chloride ingress) on the tightness of this self-healed material is analysed.

The results confirm that freeze–thaw cycles, simulating high humidity and low temperatures in cold climates, do not produce any scaling or tightness degradation of the self-healing micro-concrete stored for 28 d after cracking. When similar samples are subjected to salt spray test, to analyse the effect of chloride ingress in marine environments, saline solution is observed to produce corrosion of the external fibres, but there is no significant degradation or scaling of the specimens. Moreover, the self-healing material stored for 28 d after cracking is durable under the salt spray test conditions for a crack width of 150 μm and clearly more durable than the reference micro-concrete for 300 μm, as confirmed by backscattering electron microscopy analysis.

In conclusion, autonomous self-healing through the proposed epoxy-amine adhesive gives rise to an increase of durability of the ultra-high-strength micro-concrete under the aggressive environments considered, which implies an increase of the service life of structures in cold climates and chloride-containing environments.

Acknowledgements

Tian Run worked under funding of the Hong Kong University of Science and Technology.

Acknowledgements are also due to the Spanish government for financial support provided through the BIA2011-29234-C02-01 project, the Basque government through the PI2012-23 project and to J. A. Sanchez and the Service Unit of IETcc-CSIC for help in the experimental work.

REFERENCES

Acker P and Behloul M (2004) Ductal® technology: a large spectrum of properties, a wide range of applications. In *Proceedings of the International Symposium on Ultra High Performance Concrete* (Schmidt M, Fehling E and Geisenhanslüke C (eds)). Kassel University Press, Kassel, Germany, pp. 11–23.

Ahlborn TM, Misson DL, Peuse EJ and Gibertson CG (2008) Durability and strength characterization of ultra-high performance concrete under variable curing regimes. In *Proceedings of the 2nd International Symposium on Ultra High Performance Concrete* (Fehling E, Schmidt M and Stürwald S (eds)), Kassel University Press, Kassel, Germany, pp. 197–204.

Alkaysi M, El-Tawil S, Liu Z and Hansen W (2016) Effects of silica powder and cement type on durability of ultra high performance concrete (UHPC). *Cement and Concrete Composites* **66**: 47–56, https://doi.org/10.1016/j.cemconcomp.2015.11.005.

Andrade C, Prieto M, Tanner P, Tavares F and d'Andrea R (2013) Testing and modeling chloride penetration into concrete. *Construction and Building Materials* **39**: 9–18, https://doi.org/10.1016/j.conbuildmat.2012.08.012.

ASTM (2003) C666/C666M-03: Standard test method for resistance of concrete to rapid freezing and thawing. ASTM International, West Conshohocken, PA, USA.

Berrocal CG, Löfgren I, Lundgren K and Tang L (2015) Corrosion initiation in cracked fibre reinforced concrete: influence of crack width, fibre type and loading conditions. *Corrosion Science* **98**: 128–139, https://doi.org/10.1016/j.corsci.2015.05.021.

Bogas JA, de Brito J and Ramos D (2016) Freeze–thaw resistance of concrete produced with fine recycled concrete aggregates. *Journal of Cleaner Production* **115**: 294–306, https://doi.org/10. 1016/j.jclepro.2015.12.065.

García Calvo JL, Pérez G, Carballosa P *et al.* (2017) Development of ultra-high performance concretes with self-healing capability. *Building Materials* **138**: 306–315, http://dx.doi.org/10. 1016/j.conbuildmat.2017.02.015.

Gonzalez M, Tighe SL, Hui K, Rahman S and de Oliveira Lima A (2016) Evaluation of freeze–thaw and scaling response of nanoconcrete for Portland Cement Concrete (PCC) pavements. *Construction and Building Materials* **120**: 465–472, https://doi.org/10.1016/j.conbuildmat. 2016.05.043.

Guerrero A, García Calvo JL, Carballosa P *et al.* (2015) An innovative self-healing system in ultra-high strength concrete under freeze–thaw cycles. In *Nanotechnology in Construction. Proceedings of NICOM5* (Sobolev K and Shah SP (eds)). Springer International Publishing, Cham, Switzerland, pp. 357–362.

Hale WM, Freyne SF and Russell BW (2009) Examining the frost resistance of high performance concrete. *Construction and Building Materials* **23(2)**: 878–888.

Huang H, Ye G, Qian C and Schlangen E (2016) Self-healing in cementitious materials: materials, methods and service conditions. *Materials and Design* **92**: 499–511, https://doi.org/10.1016/j. matdes.2015.12.091.

Kim YY, Kim JM, Bang JW and Kwon SJ (2014) Effect of cover depth, w/c ratio and crack width on half cell potential in cracked concrete exposed to salt sprayed condition. *Construction and Building Materials* **54**: 636–645, https://doi.org/10.1016/j.conbuildmat.2014.01.009.

Kwon SJ, Na UJ, Park SS and Jung SH (2009) Service life prediction of concrete wharves with early-aged crack: probabilistic approach for chloride diffusion. *Structural Safety* **31(1)**: 75–83.

Lu X, Ma F, Luke A and Wang R (2015) Assessing frost resistance of concrete by impact-echo method. *Magazine of Concrete Research* **67(6)**: 317–324, http://dx.doi.org/10.1680/macr.14. 00051.

Maes M, Van Tittelboom K and De BelieN (2014) The efficiency of self-healing cementitious materials by means of encapsulated polyurethane in chloride containing environments. *Construction and Building Materials* **71**: 528–537, https://doi.org/10.1016/j.conbuildmat.2014.08. 053.

Medina C, Sanchez de Rojas MI and Frias M (2013) Freeze–thaw durability of recycled concrete containing ceramic aggregate. *Journal of Cleaner Production* **40**: 151–160, https://doi.org/10. 1016/j.jclepro.2012.08.042.

Muhammad NZ, Shafaghat A, Keyvanfar A *et al.* (2016) Tests and methods of evaluating the self-healing efficiency of concrete: a review. *Construction and Building Materials* **112**: 1123–1132, https://doi.org/10.1016/j.conbuildmat.2016.03.017.

Pentalla V (2006) Surface and internal deterioration of concrete due to saline and non-saline freeze–thaw loads. *Cement and Concrete Research* **36(5)**: 921–928.

Pérez G, Erkizia E, Gaitero JJ *et al.* (2015a) Synthesis and characterization of epoxy encapsulating silica microcapsules and amine functionalized silica nanoparticles for development of an innovative self-healing concrete. *Materials Chemistry and Physics* **165**: 39–48, https://doi. org/10.1016/j.matchemphys.2015.08.047.

Pérez G, Gaitero JJ, Erkizia E, Jimenez I and Guerrero A (2015b) Characterisation of cement pastes with innovative self-healing system based in epoxy-amine adhesive. *Cement and Concrete Composites* **60**: 55–64, https://doi.org/10.1016/j.cemconcomp.2015.03.010.

Pérez G, García Calvo JJ, Carballosa P *et al.* (2015c) Efficiency of an innovative self-healing system in ultra-high-strength concrete under salt spray test. In *Proceedings of the 10th International*

Conference on Mechanics and Physics of Creep, Shrinkage and Durability of Concrete and Concrete Structures (Hellmich C, Pichler B and Kollegger J (eds)). American Society of Civil Engineers, Reston, VA, USA, pp. 919–928.

Richard P and Cheyrezy M (1995) Composition of reactive powder concretes. *Cement and Concrete Research* **25(7)**: 1501–1511.

Richardson A, Coventry K, Edmonson V and Dias E (2016) Crumb rubber used in concrete to provide freeze–thaw protection (optimal particle size). *Journal of Cleaner Production* **112(1)**: 599–606.

Ryan PC and O'Connor A (2016) Comparing the durability of self-compacting concretes and conventionally vibrated concretes in chloride rich environments. *Construction and Building Materials* **120**: 504–513, https://doi.org/10.1016/j.conbuildmat.2016.04.089.

Sabir BB (1997) Mechanical properties and frost resistance of silica fume concrete. *Cement and Concrete Composites* **19(4)**: 285–294.

Seify M, Rahai A and Ashrafi H (2016) Prediction of chloride content in concrete using ANN and CART. *Magazine of Concrete Research* **68(21)**: 1085–1098, http://dx.doi.org/10.1680/jmacr.15.00261.

Tang W, Kardani O and Cui H (2015) Robust evaluation of self-healing efficiency in cementitious materials – a review. *Construction and Building Materials* **81**: 233–247, https://doi.org/10.1016/j.conbuildmat.2015.02.054.

Tikkanen J, Cwirzen A and Pentalla V (2015) Freeze–thaw resistance of normal strength powder concretes. *Magazine of Concrete Research* **67(2)**: 71–81, http://dx.doi.org/10.1680/macr.14.00140.

Tziviloglou E, Wiktor V, Jonkers HM and Schlangen E (2016) Bacteria-based self-healing concrete to increase liquid tightness of cracks. *Construction and Building Materials* **122**: 118–125, https://doi.org/10.1016/j.conbuildmat.2016.06.080.

UNE (Asociación Española de Normalización) (2005) EN 196-1:2005: Métodos de ensayos de cementos. Parte 1: Determinación de resistencias mecánicas. UNE, Madrid, Spain (in Spanish).

UNE (2007) EN-ISO9227:2007: Ensayos de corrosión en atmósferas artificiales. Ensayos de niebla salina (ISO 9227:2006). UNE, Madrid, Spain (in Spanish).

UNE (2008) 83982:2008: Durabilidad del hormigón. Métodos de ensayo. Determinación de la absorción de agua por capilaridad del hormigón endurecido. Método de Fagerlund. UNE, Madrid, Spain (in Spanish).

Van Tittelboom K and De Belie N (2013) Self-healing in cementitious materials – a review. *Materials* **6**: 2182–2217, http://dx.doi.org/10.3390/ma6062182.

Vegas I, Urreta J, Frias M and Garcia R (2009) Freeze-thaw resistance of blended cements containing calcined paper sludge. *Construction and Building Materials* **23(8)**: 2862–2868.

Yang K (2016) Estimates of concrete transport properties by a two pressure water test. *Magazine of Concrete Research* **68(10)**: 530–540, http://dx.doi.org/10.1680/jmacr.15.00227.

Yuan J, Liu Y, Li H and Yang C (2016) Experimental investigation of the variation of concrete pores under the action of freeze–thaw cycles. *Procedia Engineering* **161**: 583–588, https://doi.org/10.1016/j.proeng.2016.08.696.

Zhu Y, Yang Y and Yao Y (2012) Autogenous self-healing of engineered cementitious composites under freeze-thaw cycles. *Construction and Building Materials* **34**: 522–530, https://doi.org/10.1016/j.conbuildmat.2012.03.001.

Cementitious materials

Dhir and Paine
ISBN 978-0-7277-6457-7
https://doi.org/10.1680/icetsc.64577.125
ICE Publishing: All rights reserved

Chapter 8

Influence of phase change materials on temperature rise caused by hydration heat evolution of cement-based materials

C. Qian
School of Materials Science and Engineering, Southeast University, Nanjing, China

G. Gao
School of Materials Science and Engineering, Southeast University, Nanjing, China

C. Zhu
School of Materials Science and Engineering, Southeast University, Nanjing, China

Z. Guo
Faculty of Civil Engineering and Geosciences, Delft University of Technology, Delft, the Netherlands

To evaluate the influence and effect of pre-embedded phase-change materials (PCMs) on the interior temperature rise in mass concrete, a semi-adiabatic device was used to measure the temperature changes against time in cement paste, mortar and concrete that contained different contents of PCMs. The interior maximum adiabatic temperature rise of concrete with pre-embedded PCMs was calculated based on the relation between cement hydration heat, latent heat and specific heat of PCMs in different phases. The results showed that the reduction of adiabatic temperature increased and the temperature peak was delayed as the pre-embedded PCM content was increased. Consequently, the curve of the semi-adiabatic temperature became smooth. Moreover, the reduction of the interior temperature rise increased as the cement proportion in concrete increased.

Notation

C	specific heat of concrete
C_1	specific heat of solid PCM
C_1'	specific heat of liquid PCM
F	admixture dosage of concrete per unit
K	discount coefficient, for fly ash, $k = 0.25$; for slag powder, $k = 0.3$
M_c	cement dosage of concrete per unit volume
M_F	fly ash dosage of concrete per unit volume
M_p	weight of PCM
Q	hydration heat of cement per unit

q	phase transition heat of PCM
T	phase transition temperature of PCM
T_{max}	maximum adiabatic temperature rise of concrete
T'_{max}	maximum adiabatic temperature rise of concrete embedded with PCM
$T(\tau)$	adiabatic temperature rise of concrete at the age of τ days
t_0	initial temperature of concrete
ρ	density of concrete

Introduction

After concrete is cast into the mould, the hydration heat of cement raises the interior temperature of the concrete. Because concrete in different parts of a specimen is subjected to different cooling conditions, the temperature differences within the concrete specimen will increase. Tensile stress may be generated in the concrete surface layer. Since the elastic modulus and tensile strength of concrete are very low at the early age of cement hydration, tensile stress induced by temperature difference may easily exceed the tensile strength of concrete. As a result, micro-cracks and macro-cracks occur on the surface of concrete (Ole and Per, 1999; Qu, 2007). The cracks not only reduce the rigidity and integrity of a concrete structure but they also accelerate corrosion of reinforcement and carbonation. Finally, the durability of a concrete structure significantly decreases owing to the deterioration of freeze–thaw resistance, fatigue resistance, permeability resistance and water resistance (Fernando *et al.*, 2000).

Phase-change materials (PCMs) are defined substances with a high heat of fusion which, melting and solidifying at a certain temperature, are capable of storing and releasing large amounts of energy. Heat is absorbed or released when the material changes from solid to liquid and vice versa; thus PCMs are classified as latent heat storage (LHS) units. PCMs are new materials and have been developed very quickly during recent years (Pasupathy *et al.*, 2008; Wang *et al.*, 2007). They have been widely used in solar energy storage, the recycling of industrial waste heat, the thermal treatment of electronic devices, heating and air-conditioning systems, building construction and so on. Owing to their high latent heat and small temperature change in the phase-change process, PCMs could be used to control temperature change of the surrounding environment. There are exciting possibilities in improving the performance of concrete by using PCMs (Chen *et al.*, 2003; Deal and Randy, 2007).

The flow chart of using the embedded PCMs method to control the interior temperature rise in mass concrete is shown in Figure 1. PCMs are first heated to melt and are then poured into pipes or vessels with a certain shape. The PCMs are then embedded in mass concrete by way of designed methods. When the interior temperature of mass concrete reaches the PCMs' transition temperature owing to cement hydration, the PCMs will absorb the hydration heat and change from solid to liquid. As a result, the temperature peak and degree of temperature rise in the concrete will decrease significantly, and the cracks in the concrete resulting from temperature gradients will be reduced or avoided. Moreover, the liquid PCMs can be collected, so that the negative influence of direct impregnation of cementitious materials with PCMs on mechanical properties and durability can also be avoided (Hanes *et al.*, 1992; Zhang *et al.*, 2004a).

In this paper, a method of embedded PCMs is accepted. The effects of embedded PCMs on controlling interior temperature rise of cementitious materials and the relevant factors will be explored.

Figure 1 The flow chart of using PCM to control the interior temperature rise of mass concrete

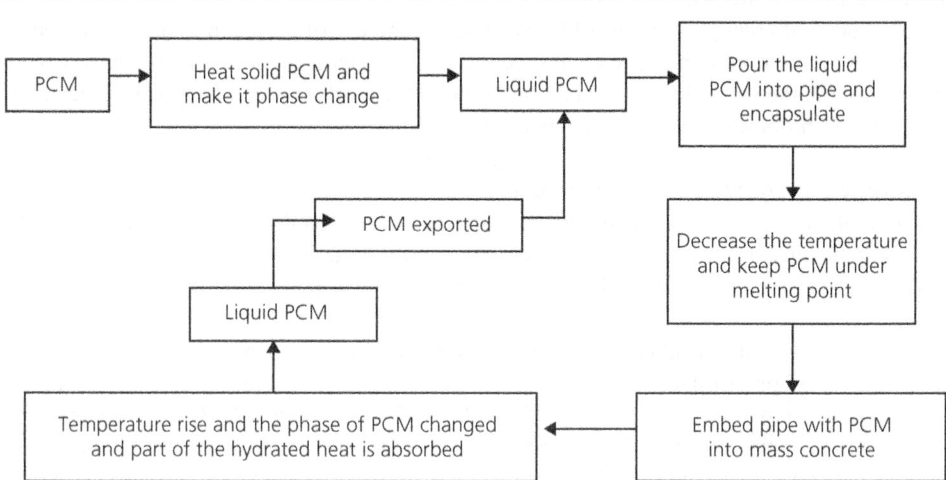

Experimental investigation

Materials

Portland cement was used in the experiments; its properties are shown in Table 1. PCM used in this study mainly consists of sodium sulfate decahydrate ($Na_2SO_4 \cdot 10H_2O$); its properties (Zhang *et al.*, 1996) are presented in Table 2.

Siliceous river sand with a fineness modulus of 2·6 and crushed rock of limestone with continuous grading ranging from 5 mm to 25 mm were used as fine and coarse aggregates. Fly ash of grade II in Chinese standard GB/T 1596-2005 was used. Its sieve residue and loss on ignition are 14·2% and 1·6% respectively.

Table 1 Basic properties of cement

Fineness 0·08 mm sieve residue %	Specific surface area: m²/kg	Water requirement of normal consistency: %	Setting time: h/min		Hydration heat: J/g	
			Initial setting	Final setting	3 days	7 days
0·4	380	28·0	1:55	2:35	221·5	288·6

Table 2 Physical properties of sodium sulfate decahydrate ($Na_2SO_4 \cdot 10H_2O$)

Molecular formula	Phase transition heat: J/g	Melting point: °C	Solid specific heat: J/g°C	Liquid specific heat: J/g°C
$Na_2SO_4 \cdot 10H_2O$	241	32·4	1·76	3·30

Experiment

Mixture proportions of cementitious materials

In this study, different proportions of PCM (0%, 3% and 6% of cement mass) were embedded in cement paste, mortar and concrete respectively. The water/cement (w/c) ratio in the cement paste was 0·3; the w/c ratio and cement/sand ratio in the mortar were 0·4 and 1/3, respectively. The mixture proportions in the concrete are shown in Table 3.

Experimental method and procedure

The hydration of cementious materials was carried out in a home-built semi-adiabatic insulated box, and the interior temperature was recorded by embedded thermocouples during the whole test. The experimental set-up is shown in Figure 2.

The inner wall of the insulated box was covered by single-layer plastic film (Maha *et al.*, 2006), and steel pipes filled with PCM were placed in the designed location during the casting process of the cementitious materials (Alferd, 2000). Thermocouples A, B and C were selected to measure the temperature in the cementitious materials (at the side and centre of the box) as well as the temperature in the PCM (Figure 2). PCMs were pre-encapsulated into the pipe of Ø31·5 × 200 mm, and the specimen dimensions were 400 × 300 × 250 mm. Thermocouple calibrators of Prova125 and T-type thermocouples with a precision of 0·2°C over the temperature range 0–120°C were used. The initial temperature of the cementitious materials was the same when the casting commenced, and the mass of the cementitious materials was also kept the same in all experiments.

Table 3 Mixture proportions in concrete: kg/m^3

Cement	Fly ash	Water	Sand	Gravel	Volume weight
330	70	182	726	1089	2400

Figure 2 The experimental device used to measure the internal temperature of cement-based materials

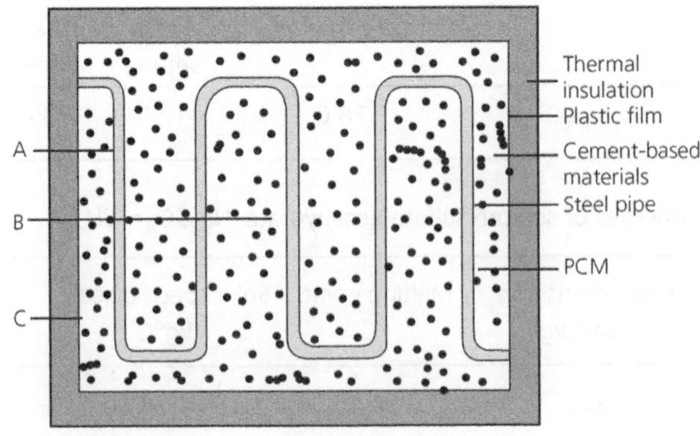

The influences of different proportions of embedded PCM and its location on the interior temperature rise were investigated.

Experimental results and discussion

Measurement of temperature started right after the cementitious materials were poured into the insulated box. The temperature was recorded under semi-adiabatic condition because of the inevitable heat emission from the insulated box during cement hydration. Experimental results are presented in Figure 3.

The influences of different embedded PCM contents in cement paste, mortar and concrete on temperature rise are first observed. Figure 3 shows that the temperature curves plotted against

Figure 3 Influence of embedded PCM on semi-adiabatic temperature rise of cementious materials: (a) cement paste; (b) mortar; (c) concrete

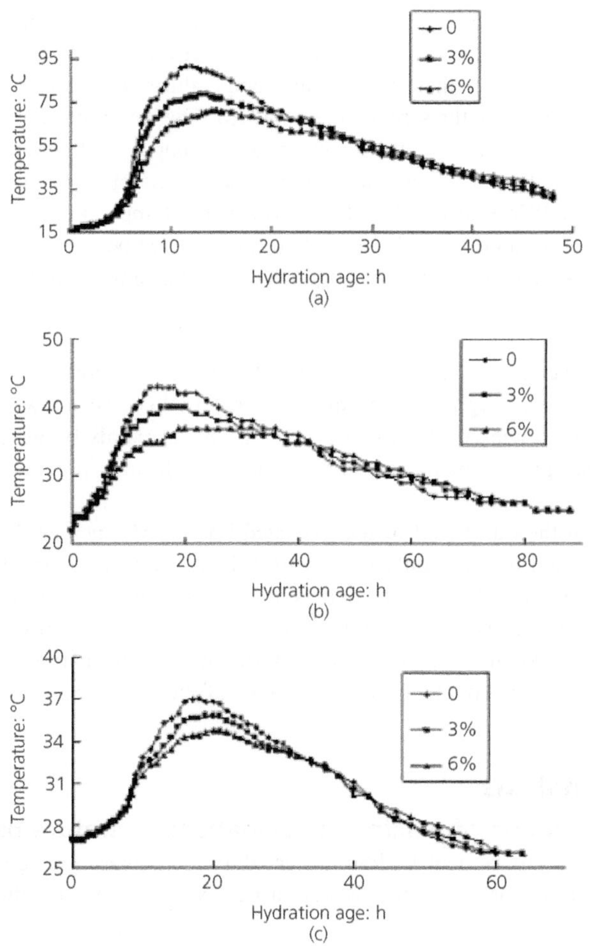

Table 4 Reduced temperature peak of samples with different PCM proportion in comparison with the sample without PCM

Material	Proportion of PCM: %	Reduced temperature peak: °C
Paste	3	13
	6	20
Mortar	3	4
	6	6
Concrete	3	1·2
	6	2·2

hydration age in all tests at a temperature of about 32°C start to deviate from each other, in other words 'shoulders' are produced on the curves at a temperature of about 32°C. This phenomenon may be explained by the fact that the transition temperature of sodium sulfate decahydrate is around 32°C and the PCM absorbs larger amounts of cement hydration heat during its phase change, so that the interior temperature rise in cementitious materials decreases in comparison with the sample without PCM. The larger the content of PCM used in the same dosage of cementitious materials, the larger the amount of hydration heat is absorbed by PCM during phase transformation, and the higher the reduction value of the temperature peak (Table 4). The PCMs also delay the arrival of the temperature peak of cementitious materials. The larger the amount of PCM used, the later the temperature peak is arrived at. The sharp curve of semi-adiabatic temperature against time tends to be smooth with the increase of PCM contents.

The interior temperature rise will decrease when the cement proportion is decreased in concrete (it also means the aggregate proportion increases). In this case, a lower relative reduction degree of temperature owing to PCM was observed. This is mainly because the aggregates in concrete also absorb part of the hydration heat during the heating process.

The temperature on the outer wall of the insulated box clearly increased during the experiments, so the insulated box can only be considered as a semi-adiabatic device. Condensed water was also observed on the inner-layer plastic film. This may be caused by the interior free water, which is evaporated when temperature increases and then condensed on the inner-layer plastic film. A large amount of heat is absorbed during the transformation of water from liquid to gas. It can therefore be concluded that only part of the hydration heat raises the interior temperature.

Theoretical analysis

Theoretical calculation of maximum adiabatic temperature rise

PCMs could absorb part cement hydration heat during its phase change, and the solid or liquid PCMs could also absorb some hydration here when the temperature rises (Prashant *et al.*, 2008).

Cement hydration heat will be completely transformed into interior temperature rise of cementitious materials without PCMs under adiabatic condition. The diabatic temperature rise $T(\tau)$ of concrete can be theoretically calculated by (Wang and Zhou, 2006; Zhang *et al.*, 2004b)

$$T(\tau) = \frac{(M_C + KM_F)Q}{C\rho}(1 - e^{-m\tau})$$

(1)

where M_C and M_F are masses of cement and admixture per unit of concrete volume, Q is hydration heat per unit of cement mass; K is reduction coefficient, $K = 0\cdot25$ for fly ash and $K = 0\cdot3$ for mineral powder; C is the specific heat of cement; ρ is density of concrete; τ is the hydration age; and m is a constant related to the initial temperature of the concrete.

If the degree of hydration is high enough, the maximum adiabatic temperature rise T_{max} of the cementitious materials could be expressed by

$$T_{max} = \frac{(M_C + KM_F)Q}{C\rho}$$

(2)

When PCMs were pre-embedded in the concrete, solid and liquid PCMs will also absorb heat before and after phase transition. The adiabatic temperature rise can therefore be divided into three periods. In the first period the interior temperature rise below the phase transition temperature and PCM exists in a solid state; the heat absorbed by PCM in this period is $Q_1 = C_1 M_p (T - t_0)$ where C_1 and M_p are the specific heat and mass of solid PCM respectively and t_0 is the initial temperature of the concrete. When the interior temperature is equal to the phase transition temperature PCM changes from solid to liquid. This is the second period and the heat absorbed by PCM is $Q_2 = M_p q$ where q is the transition heat of PCM. In the third period the interior temperature rise above the phase transition temperature and PCM exists in liquid state. The heat absorbed by PCM in this period is $Q_3 = C_1' M_p(T_{max}' - T)$ where C_1' is the specific heat of liquid PCM and T_{max}' is the maximum adiabatic temperature rise with embedded PCM. The total heat absorbed by PCM during the heating process can therefore be summarised as

$$Q' = C_1 M_p(T - t_0) + M_p q + C_1' M_p(T_{max}' - T)$$

(3)

Because the heat released from cementitious materials was partly absorbed by PCM, the interior temperature rise of cementitious materials could not reach the value of T_{max}. The theoretical calculation of maximum adiabatic temperature rise of concrete with pre-embedded PCMs could be described by

$$T_{max}' = \frac{(M_c + KM_F)Q - \left[C_1 M_p(T - t_0) + q M_p - C_1' M_p(T_{max}' - T)\right]}{C\rho}$$

(4)

or

$$T'_{\max} = \frac{(M_c + K M_F)Q - M_p[C_1(T - t_0) + q - C'_1 T]}{C\rho + C'_1 M_p} \tag{5}$$

Equation 5 demonstrates that the maximum adiabatic temperature rise of the sample with PCM depends on the proportion and hydration heat of cement; the proportion, specific heat, phase transition heat and phase transition temperature of PCM; and the specific heat, density and initial temperature of concrete.

According to ASTM C 150 standard, hydration heat of Portland cement is about 500 J/g when the cement completely hydrated. The cement hydration degree in cement paste at the age of 180 days varies from 64·39% to 81·13% with a w/c ratio of 0·3 to 0·5 (Zhang *et al.*, 2006). It is therefore assumed that the final hydration degree is 80% and the relevant hydration heat is 400 J/g. The density and specific heat of concrete are assumed to be 2400 kg/m^3 and 0·96 J/g°C (Chen *et al.*, 2007). The initial temperature is 25°C. The influence of cement and PCM dosages on the maximum adiabatic temperature rise based on Equation 5 can be calculated as shown in Figure 4. It is observed that the reduction value of the maximum adiabatic temperature rise of concrete rises as the content of pre-embedded PCM increases and the value of the maximum adiabatic temperature rise increases as the cement proportion in concrete increases. These conclusions from numerical calculation are consistent with experimental results.

Reduction degree of maximum adiabatic temperature rise

For a certain dosage of cement, the degree of reduction of the maximum adiabatic temperature rise owing to the change of PCM dosage can be calculated based on Equations 2 and 5

$$\alpha = \left(1 - \frac{T'_{\max}}{T_{\max}}\right) \times 100\% \tag{6}$$

Figure 4 The maximum adiabatic temperature rise at PCM content from 0 to 30%

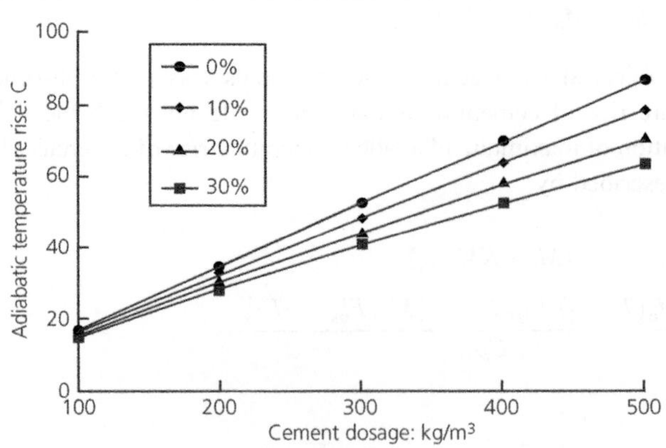

Substituting Equations 2 and 5 into Equation 6 yields

$$\alpha = \left\{ 1 - \frac{\dfrac{C\rho Q(M_\mathrm{c} + KM_\mathrm{F})}{-C\rho M_\mathrm{p}[C_1(T - t_0) + q - C_1'T]}}{(C\rho + C_1 M_\mathrm{p})(M_\mathrm{c} + KM_\mathrm{F})Q} \right\} \times 100\% \tag{7}$$

Figure 5 illustrates the influence of PCM and cement proportions on the degree of reduction of the maximum adiabatic temperature rise. The reduction of the maximum adiabatic temperature rise increases with the increasing value of PCM because more hydration heat is absorbed by PCM. Furthermore, the higher the dosage of cement contained in unit volume of concrete, the higher the value of the maximum adiabatic temperature rise will be. For example, when the cement dosage is 400 kg/m^3 with 10% of PCM, the maximum adiabatic temperature rise is 6·17°C and the reduction degree is 8·9%; when the cement dosage is 100 kg/m^3 with 10% of PCM, the maximum adiabatic temperature rise is 0·87°C and the reduction degree is 5%. Therefore, if the cement dosage is lower, both absolute value and reduction degree of adiabatic temperature rise are lower.

It can be concluded that PCM can be used to reduce adiabatic temperature rise in at least two aspects. One is increasing the embedded PCM dosage combined with proper selection of an embedding method and another is that PCM can be applied to the materials with high cement dosage. Reducing the interior temperature rise of cementitious materials with embedded PCM is a brand-new method, so many details, such as embedding method, compatibility between PCM and encapsulation vessels, influences on performance of concrete and structure, recycling of liquid PCM and so on all require further investigation.

Figure 5 Calculated relationship between content of PCM and reduction extent of maximum adiabatic temperature rise

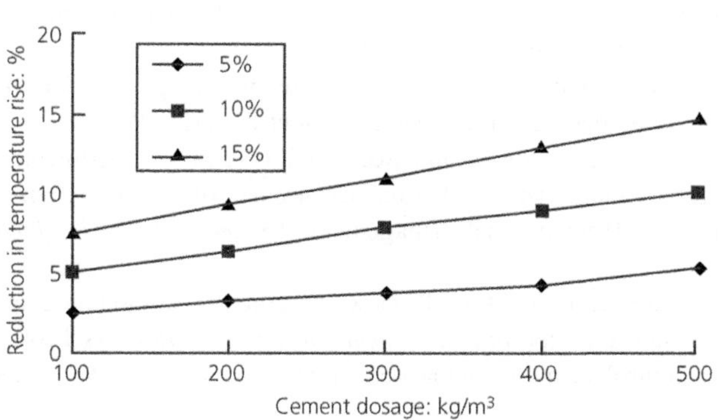

Conclusions

(*a*) PCMs pre-embedded in cementitious materials reduce the interior temperature rise caused by hydration heat, and delay the arrival of temperature peak. The curve of interior temperature rise tends to slow down with the increasing PCM dosage.

(*b*) The maximum adiabatic temperature rise and the degree of reduction of maximum adiabatic temperature rise for cementitious materials with PCM are theoretically calculated.

(*c*) The degree of reduction of adiabatic temperature rise of cementitious materials has a close relationship with PCM dosage and cement dosage per unit volume of concrete. The higher the proportion of PCM or the higher the proportion of cement, the larger the reduction ratio of the adiabatic temperature rise will be.

Acknowledgements

The authors greatly appreciate the financial support from the National Basic Research Programme of China (No. 2009CB623200), the National Natural Science Foundation of China (No. 50539040) and the Scientific Research Foundation of Graduate School of Southeast University (No. YBJJ 0725).

REFERENCES

Alferd H (2000) 15 years of R&D in central solar heating in DENMARK. *Solar Energy* **69(6)**: 437–447.

Chen DP, Qian CX, Wang H and Liu JH (2007) Research on determination and calculation method of specific heat capacity of cement-based materials. *Journal of Building Materials* **10(2)**: 127–131.

Chen MZ, He Z and Liang WQ (2003) Study on temperature self-control in concrete with phase change materials. *Journal of Building Materials* **6(4)**: 386–390.

Deal PB and Randy T (2007) Potential applications of phase change materials in concrete technology. *Cement and Concrete Composites* **29(7)**: 527–532.

Fernando H, Guillermo S and Juan AB (2000) Temperature and cracking control of the deck slab concrete at early ages. *Automated Construction* **9(5–6)**: 437–445.

Hanes DW, Banu D and Feldman D (1992) The stability of phase change materials in concrete. *Solar Energy Materials and Solar Cells* **27(2)**: 103–118.

Maha A, Andre B, Hebert S and Danlel Q (2006) Thermal testing and numerical simulation of a prototype cell using light wallboards coupling vacuum isolation panels and phase change material. *Energy and Building* **38(6)**: 673–681.

Ole MJ and Per FH (1999) Influence of temperature on autogenous deformation and relative humidity change in hardening cement paste. *Cement Concrete Research* **34(4)**: 567–575.

Pasupathy A, Athanasius L, Velraj R and Seeniral RV (2008) Experimental investigation and numerical simulation analysis on the thermal performance of a building roof incorporating phase change material (PCM) for thermal management. *Applied Thermal Engineering* **5(5–6)**: 556–565.

Prashant V, Varum and Singal SK (2008) Review of mathematical modeling on latent heat thermal energy storage systems. *Renewable and Sustainable Energy Reviews* **12(4)**: 999–1031.

Qu W (2007) Crack control technique of mass concrete construction. *Building Science* **23(5)**: 75–77.

Wang DD and Zhou SQ (2006) Experimental study of temperature rising of mass concrete insulating. *Coal Ash China* **5(4)**: 3–6.

Wang SX, Du SF, Weng XL and Deng LJ (2007) Studied for epikote-sodiam dihydrogen phosphate complex material of phase change energy storage. *Journal of Functional Materials* **38(4)**: 646–647.

Zhang D, Li ZJ, Zhou JM and Wu KR (2004a) Development of thermal energy storage concrete. *Cement and Concrete Research* **34(6)**: 927–934.

Zhang YP, Hu HP, Kong XD and Su YY (1996) *Theory and Application of Energy Storage with Phase Change Materials*. Publishing House of the Science and Technology University of China, p. 341.

Zhang YS, Sun W, Zheng KR and Jia YT (2006) Hydration process of Portland cement–fly ash pastes. *Journal of Southeast University* **36(1)**: 118–123.

Zhang ZM, Feng SR, Shi QC and Wang JH (2004b) Adiabatic temperature rise of concrete based on equivalent time. *Journal of Hehai University* **32(5)**: 573–577.

Wang PD and Zhang LY (2006) Temperature rise analysis [...] computing [...] boundary of [...]. *Journal of [...]*, 31(3): 1–5.

Wang SX, Du M, Li XZ and Yang LX (2007) Selected forms [...] and insulation properties of [...] of [...] epoxy [...]. *Chinese [...] Journal*, 27(4): [...].

Zhang QG, Li XM, Ma JM and Wu XL (2010) Temperature of thermal stress voltage compose [...] and [...] film. *Transactions of [...]*, 25(9): 62–65.

Zhang YR, Ho HP, Kang SD and Su YS (1994) [...] rate and temperature [...]. [...], Metallurgical Industry Press House. Science and Technology University Press [...].

Zhou JL and Ma JC (2007) [...] distribution in core of insulated current [...] with [...]. *Journal of [...]*, 36(3): 15–19.

Zhou M, Chen X, Shen K and Wang YF (2008) Ambient temperature rise of [...] cables based on [...]. *Journal of [...]*, 30(5): [...].

Dhir and Paine
ISBN 978-0-7277-6457-7
https://doi.org/10.1680/icetsc.64577.137
ICE Publishing: All rights reserved

Chapter 9

Dispersion effectiveness of carbon nanotubes in smart cementitious materials

Leonard Chia
Department of Civil and Environmental Engineering, North Dakota State University, Fargo, ND, USA

Ying Huang
Department of Civil and Environmental Engineering, North Dakota State University, Fargo, ND, USA (corresponding author: ying.huang@ndsu.edu)

Applying carbon nanotubes (CNTs) as an additive in cement has high potential to achieve smart cementitious materials. However, the small fraction of CNTs in the cementitious materials may not be sufficient to provide strong piezo-resistive signals for monitoring the materials' performance accurately in the field. In this study, a theoretical transfer function is developed to analyse the dispersing effectiveness of CNTs in order to search for a stronger piezo-resistance effect in smart cementitious materials. The theoretical transfer function model introduces a dispersing effectiveness coefficient into the intrinsic resistance and the inter-tube resistances of CNTs in the cementitious mix matrix. To validate the developed model, two different dispersing methods, including direct mixing and surfactant methods, were investigated through laboratory experiments. The experimental data showed that the developed transfer function can distinguish the dispersing effectiveness well for the two different methods investigated. This model can be used to guide evaluation regarding the quality control of CNT-based smart cementitious materials and investigation of the effectiveness of new developments of dispersion methods in the future.

Notation

C_A	contributing factor from percentage of carbon nanotubes (CNTs) by weight used in the cement
C_B	influence factor to consider the dispersing effectiveness
K_B	final influence factor from the dispersing effectiveness
k_1	strain sensitivity of the intrinsic resistance of CNTs to resistance changes
L	carbon nanotube length with strain
L_0	initial length of CNT
R_0	base resistance of CNTs

s	gap between nanotubes
s_0	initial gap between nanotubes
ΔR^{T}	tunnelling resistance change
ΔR_{tube}	intrinsic resistance changes of the CNTs
ε	strain
λ	constant related to average height of potential barrier
ρ	constant equal to resistance change of tube per unit length

Introduction

The application of cementitious materials has become popular as a base material for various civil engineering materials such as concrete and asphalt, which are widely used as structural materials. The onset of local damage in structures with cementitious materials, such as concentrated stresses or strains, cracking and delamination, is often difficult to detect, and such damage has a long-term implication regarding the performance of these composite structures (Thostenson and Chou, 2008). To detect such instances of local damage and monitor the structural health conditions of structures made from cementitious materials, currently, several non-destructive detection techniques can be considered, such as acoustic, ultrasonic, X-ray or eddy current inspection (Ativitavas *et al.*, 2006; Dzenis and Saunders, 2002), as well as embedded sensors in structures (Farrar and Worden, 2007; Giurgiutiu *et al.*, 2002; Zou *et al.*, 2000). However, the use of non-destructive inspection tools may require intensive labour and time from experienced inspection engineers, or sometimes even the disassembly of the structural components. In recent years, to achieve real-time structural health monitoring in structures, smart materials such as piezo-resistive cementitious materials modified by carbon nanotube (CNT) additives may represent an alternative approach.

A piezo-resistive smart material changes its electrical resistance due to mechanical stresses or strains. Materials that are currently known to have piezo-resistive effects include germanium, polycrystalline silicon, amorphous silicon, silicon carbide, CNTs, graphite and steel fibres (Lyshevski, 2002). The piezo-resistive effects have been popularly applied in civil engineering fields to make strain gauges that can measure strains on structures (Kon *et al.*, 2007). Among the piezo-resistive materials, CNTs, graphite and steel fibres have been extensively investigated for use in smart cementitious materials such as smart concrete.

Carbon nanotubes, owing to the fact that a small volume of additives can enable a strong piezo-resistive effect, have been considered a promising candidate to enable the self-sensing capability of cementitious materials. A CNT is an allotrope of carbon sharing a cylindrical nanostructure. The CNTs have a long, hollow structure, with walls formed by sheets of carbon just one atom thick, called graphene. These graphene sheets are rolled at specific and discrete angles, using a combination of different rolling angle and radius. Usually, the length to diameter ratio of the graphene sheets making CNTs is smaller than 132 000 000:1, which enables extraordinary strength and unique electrical properties (Wang *et al.*, 2009). These unique properties make them potentially useful in a wide variety of applications in nanotechnology, electronics, optics and other fields of materials science (Naidu *et al.*, 2014; Saito, 2010). Based on the numbers of graphene sheets in CNTs, the CNTs can be either single-walled nanotubes (SWNTs) or multi-walled nanotubes (MWNTs) (Sharma *et al.*, 2015). SWNTs are hollow, single cylinders of a graphene sheet, which are defined by their diameter and their chirality. The diameter of SWNTs varies from 0·5 nm to 5 nm. Depending on the chirality, SWNTs may either be metallic or semiconducting (Maiti *et al.*, 2002). MWNTs are a group of concentric SWNTs, capped at both ends, with diameters in the range from several nanometres up to 200 nm. These concentric nanotubes are held together by van der Waals bonding. MWNTs form complex systems with different wall numbers, structures and properties (Ganesh, 2013). When CNTs are subjected to strain, changes in the conductive network of the CNTs arises, which results in an increase in their resistivity. With the change of resistivity, the electrical resistance in the CNTs also changes, which can be used as a 'smart' function of the added materials.

It has long been assumed that CNTs were effectively dispersed in water through direct mixing with water. Recently, however, studies have shown that this assumption that the direct mixing of CNTs in water, without any dispersing methods, will effectively disperse the CNT particles into the cement, may not be entirely valid. Direct mixing of the CNTs into water without any dispersion methods would result in small piezo-resistive changes to dynamic loadings (Materazzzi *et al.*, 2013). In the case when no dispersion methods are used, the CNTs are more dispersed at a lower water–cement (w/c) ratio. Also, the variation of piezo-resistive sensitivity and stability induced by the water content decreases with lower w/c ratio (Kim *et al.*, 2014). These results have shown that the various dispersing methods matter in terms of piezo-resistive effects (Materazzzi *et al.*, 2013) and more investigations are needed. More recently, the piezo-resistive response of two different dispersion methods of CNTs in cement have been studied, including the use of acid surface treatment and surface modification by the surfactant sodium dodecylbenzene sulfonate (NaDDBS). The experimental results showed that the acid treatment method had a much stronger and more accurate response compared to the surfactant wrapping method (Han *et al.*, 2014; Kim *et al.*, 2014). When using the acid treatment method, although it provided a stronger piezo-resistive response, the use of acid treatment was difficult to scale up for larger samples. The use of strong acid also made the treatment difficult to implement, as it would pose danger in the field. Although not as good as the acid treatment method, the piezo-resistive response from the surfactant dispersion method was also found to be promising (Frank *et al.*, 1998). Further studies performed on two different surfactants for surface modifications – sodium dodecyl sulfate (SDS) and NaDDBS – also showed that NaDDBS is more stable and sensitive to the external force compared with SDS when dispersing the CNTs (Yu and Kwon, 2009). In addition, superplasticiser and silica fume have been used as the surfactant to mix CNTs into cement; however, the researchers claimed that these methods were not effective (Yu and Kwon, 2012). In all of these investigations, MWNTs were adopted, as they are more sensitive to stress changes compared with SWNTs.

In current practice using CNTs in smart cement materials, when samples or structural components are made using certain dispersing methods – either acid treatment or the surfactants method – it is assumed that the dispersing has been effective and the data is valid. There has been no quality control to ensure that the required strain measurement resolution and sensitivity has been obtained from the smart cementitious materials. However, as the dispersion might not be uniform or effective, the actual strain or stress measurement accuracy of the smart cementitious materials could vary significantly. Thus, before the application of CNT smart cement, an investigation is needed as to whether measurement made using these materials are valid. To date, however, there has been no theoretical transfer function model that can easily identify and evaluate the effectiveness of the dispersing methods; this is therefore an urgent need for the application of CNT-enabled smart cementitious materials.

This paper develops a theoretical transfer function to investigate the effectiveness of dispersing methods for dispersing CNTs into cementitious materials and validates the model through laboratory experiments. The developed transfer function model will provide evaluation guidance on the investigation to determine the sensing sensitivity, resolution, accuracy and other sensing characteristics of smart cementitious materials and to further determine whether the dispersion is the cause if the materials do not meet the required resolution for a quality CNT-enabled smart cementitious material. The rest of the paper is organised as follows: the section entitled 'General theoretical analysis on dispersion effectiveness of CNTs' develops the theoretical algorithm; the section entitled 'Methodology' prepares the methodology for the laboratory validation experiments; the section entitled 'Experimental results and discussion' validates the theoretical analysis using experiments; and the final section provides the conclusions and outlines future work.

General theoretical analysis on dispersion effectiveness of CNTs

The CNT-modified smart cementitious materials use monitored strains to assess structural behaviour. The strain-sensing phenomenon in a CNTs network is attributed to two types of resistance: the intrinsic resistance, R_{tube}, and the intertube resistances, R_{Inter}. In the following sections, detail relations between the strains to the two types of resistances of CNTs are derived, so that the transfer function of strain changes to resistances changes can be developed.

Intrinsic resistance of CNTs

Carbon nanotubes can act as good conductors because of their one-dimensional structures, which allow electronic transport to occur ballistically (Azhari and Banthia, 2012; Han and Yu, 2014). MWNTs have been found to have an intrinsic resistance of $0 \cdot 2$–$0 \cdot 4 \, \text{k}\Omega/\mu\text{m}$ (Berger *et al.*, 2002; Liang *et al.*, 2001). The intrinsic resistance is subjected to modification under strain. Studies have shown that the intrinsic resistance of SWNTs increases considerably at a relatively small strain (Heyd *et al.*, 1997; Zamkov *et al.*, 2006). The intrinsic resistance of CNTs increases proportionally with the applied strain with a dependence on the chiral angle of the CNTs (Yang and Han, 2000), which can be expressed as below

$$\Delta R_{\text{tube}} = (L_0 - L)\rho = G_{\text{tube}}^{-1} \tag{1}$$

where ΔR_{tube} denotes the intrinsic resistance changes of the CNTs, L_0 is the initial length of the tube, L is the tube length with strain, and ρ is a constant which equals the resistance change of the tube per unit length.

Thus, the intrinsic resistance changes of CNTs with the strain, ε, can be derived as

$$\Delta R_{\text{tube}} = \frac{(L_0 - L)}{L}\rho L = \rho L \varepsilon = k_1 \varepsilon \tag{2}$$

in which k_1 is the strain sensitivity of the intrinsic resistance of CNTs to the resistance changes, and is equal to ρL.

140

Intertube resistance of CNTs

Despite the intrinsically high conductivity of CNTs, conduction in a CNT network is not correspondingly efficient. This is because intertube resistance is comparatively higher than the intrinsic resistance. The intertube resistances include the contact resistance, R^C, which is the resistance between tubes that are physically in contact, and the tunnelling resistance, R^T, which is the resistance between tubes that are separated by a small gap.

The contact resistance between nanotubes in CNTs is introduced by physical contact between tubes, such that the conduction takes place between these CNTs through electron diffusion. The contact resistance has been shown to depend greatly on the contact region and has large values varying from a few hundreds to a thousand kΩ. These values depend on factors such as the arrangement of molecules across the interface and the extent of the interfacial surface (Minot *et al.*, 2003). When the arrangement of molecules and the extent of the interfacial surface are settled, the contact resistance of CNTs remains stable in most cases, resulting in insensitivity to the external strains placed upon them. Therefore, when measuring intertube resistance changes of CNTs with various strains, the major contribution is assumed to come from the changes of the tunnelling resistance, R^T, and the changes in contact resistance are neglected.

The tunnelling resistance of the CNTs, R^T, is introduced by conduction between the gaps of the CNTs. Tunnelling resistance can be calculated as below (Yang *et al.*, 1999)

$$R^T = R_0 e^{\lambda s} \tag{3}$$

where s is the gap in between the nanotubes, R_0 is the base resistance and λ is a constant related to the average height of the potential barrier, the values of R_0 and λ can be obtained as below

$$R_0 = \frac{1}{C_1} \frac{s_0}{\sqrt{K}} \quad \text{and} \quad \lambda = C_2 \sqrt{K} \tag{4}$$

in which s_0 is the initial gap in between the nanotubes; C_1 and C_2 are constants with a value of $3 \cdot 16 \times 10^{10}$ and $1 \cdot 0125$, respectively; and K is equal to the average height of the potential barrier, which is assumed to be a constant of 6.

If strain, ε, occurs on the CNTs, it will introduce gap distance changes in between the nanotubes, from s_0, the initial gap between the nanotubes, to s, the current gap distance. Thus, based on Equation 3, the tunnelling resistance changes, ΔR^T, can be calculated as

$$\begin{aligned} \Delta R^T &= R_0 e^{\lambda s_0} - R_0 e^{\lambda s} = R_0 e^{\lambda s_0} \left(1 - e^{\lambda(s_0 - s)}\right) \\ &= R_0 e^{\lambda s_0} \left(1 - e^{\lambda s_0}(s_0 - s)/s_0\right) = R_0 e^{\lambda s_0} \left(1 - e^{\lambda s_0 \varepsilon}\right) \end{aligned} \tag{5}$$

Thus, if $k_2 = \lambda s_0$ and $K_B = R_0 e^{k_2}$ are assigned, Equation 5 can be rewritten as

$$\Delta R^T = R_0 e^{k_2} \left(1 - e^{k_2 s}\right) = K_B \left(1 - e^{k_2 s}\right) \tag{6}$$

Total resistance from CNTs

The total resistances, R_{Total}, induced by the CNTs, therefore, can be derived as the summation of the intrinsic resistance of CNTs and the intertube resistance of CNTs

$$\Delta R_{\text{Total}} = \Delta R_{\text{Tube}} + \Delta R^{\text{T}} = k_1\varepsilon + R_0 e^{k_2}\left(1 - e^{k_2}\varepsilon\right)$$
$$= K_{\text{B}} + k_1\varepsilon - K_{\text{B}} e^{k_2\varepsilon} \tag{7}$$

It is obvious from Equation 7 that the piezo-resistance response of the CNTs to the strains on them is a non-linear behaviour. However, with laboratory calibration, k_1, k_2 and K_{B} can be obtained and the resistance changes of the CNTs with strains can be derived theoretically.

Piezo-resistance of CNT-modified cementitious materials with dispersing effectiveness considered

If 100% of CNTs are used as a sensing material, Equation 7 can be used to calculate the resistance change of the CNTs for corresponding strains on them, which is the sensing principle for piezo-resistive sensors made by pure CNTs. However, for practical application, sensors based on pure CNTs are very expensive, which limits the size of the sensors. For smart structural material, such as smart cementitious materials, it is only affordable to add a small portion of CNTs into the base cementitious material for modification of the material's properties and thereby enabling its sensing capability.

Thus, the theoretical piezo-resistance response of the CNTs that are incorporated with the cementitious materials may vary from Equation 7. However, Equation 7 can still be used as a base equation to calculate the piezo-resistance of individual CNTs in the cementitious materials. In this study, the hypothesis is that the piezo-resistance of CNT-modified cementitious materials toward strains, ε, still remains a function of the total resistance of the individual CNTs, ΔR_{total}, with dependence on the percentage of CNTs used for the cement mortar (%CNT), the w/c ratio and the dispersion effectiveness of the CNTs in the cementitious materials. The %CNT will influence both the intrinsic and intertube resistances of CNTs in cement mortar. If the same percentage of CNTs is used, the intrinsic resistance of the CNTs will not change much with the dispersing effectiveness contributed by the w/c and dispersing methods. The dispersing effectiveness will have the major influence on the intertube resistance, as it changes the distance in between the nanotubes, which is the tunneling resistance, which in turn will significantly change the parameter R_0, and thus, K_{B} in Equation 7. Based on this hypothesis, the transfer function on the piezo-resistance change of CNT-modified cementitious materials ΔR to strain changes can then be expressed as follows

$$\Delta R = C_{\text{A}} K_{\text{B}} + k_1\varepsilon - K_{\text{B}} e^{k_2\varepsilon} \tag{8}$$

where C_{A} is a contributing factor from the percentage of CNTs by weight used in the cementitious materials, %CNT; K_{B} is the influencing factor to consider the dispersing effectiveness, which is contributed by two major factors: the w/c ratio of the cementitious material and the dispersion method used for the CNTs. Thus, if 100% of CNTs is used, C_{A} will be equal to 1. Also if the CNTs dispersed uniformly and effectively in the cementitious materials, $R_0 = 1$ in $K_B = R_0 e^{k_2}$. In this ultimate circumstance, the resistance will be the total resistance of the CNTs. As previous studies (Azhari and Banthia, 2012) have been intensively performed on the influences of percentage CNTs and w/c ratios on the piezo-resistance of CNT-modified cementitious materials, in this study, the authors focus on the contributing factor from the effectiveness of various dispersing methods, K_B. To investigate the contributing factor of the dispersing effectiveness from various dispersing methods, all of the samples in the present study are maintained at the same percentage of CNTs and the same w/c ratio.

Equation 8 can serve as a transfer function between the piezo-resistance and the strain changes of the smart cementitious materials. This transfer function can provide guidance when determining the strain measurement resolution, sensitivity, dynamic range, accuracy and other sensing characteristics. These sensing characteristics can be used to predict and evaluate the performance of the smart cement as well as providing guidance in the investigation of the CNTs' dispersion. In the developed theoretical transfer function model, if the dispersion factor (K_B) in Equation 8 of one specific dispersion method is calibrated using laboratory calibration tests, the strain resolution of the smart cement made by this dispersion method can be predicted using the condition that the measured resistance change needs to be larger than zero to be effective, which is $\Delta R > 0$. Conversely, the minimum required dispersion factor can be calculated using Equation 8 if a given minimum strain, also known as resolution, is desired. For development of new dispersion methods, this guidance is essential to save time and effort, and thereby avoid wasting extensive amounts of time in conducting the experiments.

Methodology

To validate the theoretical transfer function developed in the previous section and to approve the effectiveness of Equation 8, in this paper, two dispersion methods were investigated in the laboratory: the direct mixing method, without any specific dispersing approaches considered, and the surfactant method. As seen in Equation 8, the piezo-resistance change of the CNT-modified cementitious materials ΔR considers all three influencing factors, with C_A for %CNT, k_1 for the strain sensitivity of the intrinsic resistance, and K_B representing the combined effects of the w/c ratio and the dispersion method of the CNTs. As stated previously, since the percentage of CNTs and w/c ratios have been studied previously (Azhari and Banthia, 2012), in this study, the focus is on the CNTs dispersion effect in K_B. The CNTs used in this study are MWNTs with length $<100\,\text{nm}$, supplied by SkySpring Nanomaterials, Inc., and the cement used is Type I Portland cement supplied by Holcim Inc., USA. Table 1 presents the material properties of the CNTs used.

For all the CNT-modified cement mortar samples to be prepared for investigations on the effectiveness of the various dispersing methods, a w/c ratio of 0·6 and a percentage of CNTs of 0·1% by weight were used. The size of the samples prepared was 2 in. by 2 in. (5·08 cm by 5·08 cm) cubes. Thus, each sample contained 400 g of cement, 240 ml of water and 0·4 g of CNTs. All the samples were prepared at room temperature (22°C±2°C).

Table 1 Properties of multi-wall CNTs

Parameters	Values
Outside diameter	50–100 nm
Inside diameter	5–10 nm
Length	5–20 μm
Purity	>95 weight %
Ash	<1·5 weight %
Special surface area	>60 m²/g
Amorphous carbon	<3·0%
Bulk density	0·28 g/cm³
True density	~2·1 g/cm³

Direct mixing method

The direct mixing method is a common practice for mixing CNTs into cement mortar. This method directly mixes the CNTs with cement and water, without any treatment of the CNTs. There are three steps to prepare a cubic sample using the direct mixing method, as follows.

(*a*) The 0·4 g CNTs is directly mixed with 240 ml water to create a solution.
(*b*) The solution is fully mixed using a magnetism stirrer for 3 min; the fully mixed solution is shown in Figure 1 on the right.
(*c*) The solution is further used to mix with 400 g of cement to make a cubic sample.

Figure 1 CNTs in water using direct mixing method on the right compared with NaDDBS treated CNTs solution on the left

Surfactant method using NaDDBS

The surfactant dispersion method was also investigated in this study by applying NaDDBS for surface modification of the CNTs. The NaDDBS used was supplied by Sigma-Aldrich Co., USA. A critical micelle concentration of NaDDBS in water, $1·4 \times 10^{-2}$ mol/l, was taken as the input surfactant concentration. There are four steps to prepare the solution of surfactant-treated CNTs solution.

(a) $1·17$ g of NaDDBS is mixed with 240 ml of water using a magnetism stirrer for approximately 3 min, as shown in Figure 2(a).

(b) $0·4$ g of CNTs are added into this aqueous solution and sonicated using a sonicator for 2 h, as shown in Figure 2(b), to make a uniformly dispersed suspension, as shown in Figure 1 on the left; the colour of the NaDDBS-treated liquid is much darker than that produced by the direct mixing method when comparing the two solutions.

(c) The NaDDBS-treated CNTs solution is then mixed with 400 g of cement for about 3 min.

(d) In order to decrease the air bubbles in the CNTs filled with cementitious materials, as caused by the NaDDBS, $0·25\%$ of defoamer by volume is added and mixed with the modified cement for another 3 min. The defoamer used in this study is tributyl phosphate, supplied by Sigma-Aldrich Co., USA.

After dispersing the CNTs into water using the two methods described above, the solutions were further mixed with cement and placed into 2 in. cubic moulds in order to prepare the

Figure 2 (a) Mixing NaDDBS in water and (b) mixing CNTs with NaDDBS solution

(a)   (b)

cubic samples, as shown in Figure 3(a). To perform measurements of the piezo-resistive effects, two electrical wires were placed 0·5 in. deep into the samples and 0·5 in. apart from each other. The samples were demoulded in 24 h and left to cure in water for 7 d; the demoulded samples are show in Figure 3(b). The samples were then left to air dry at a room temperature of 22°C ± 2°C for 10 d. Three samples were fabricated for each dispersion method, in addition to three control samples without any nanotubes.

Figure 3 (a) The 2 inch (5·08 cm) cubic mould for sample fabrication and (b) sample with electric wires

(a)

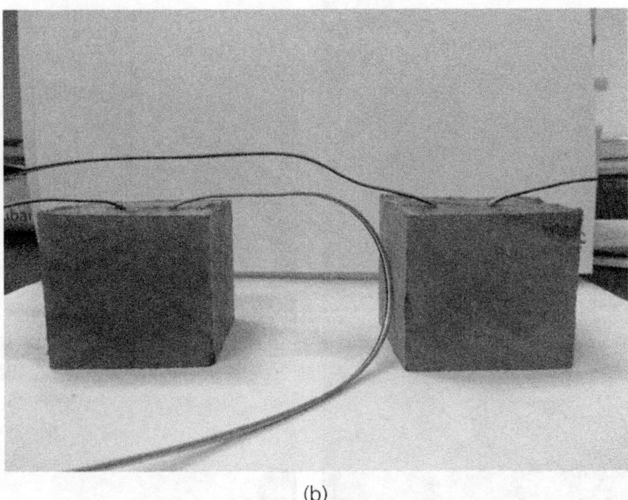

(b)

Experimental results and discussion

With the samples prepared, laboratory experiments on the piezo-resistive responses of the CNT-modified cement mortars prepared by the two dispersing methods were performed; the test results were analysed to validate the developed theoretical model derived in the earlier section entitled 'General theoretical analysis on dispersion effectiveness of CNTs'.

Experimental set-up

Dynamic loading tests were performed on the prepared samples. Figures 4(a) and 4(b) illustrate the laboratory test set-up. Compressive loads were applied using a material testing machine (MTS 858, Materials Testing Systems, Inc., USA). The electrical resistance changes of the CNT-modified cement mortar samples were measured in the direction of the compressive stress, which was perpendicular to the electrodes under repeated compressive loading. The electrical resistance responses were measured using a two-electrode method, employing a digital bench multi-meter (BK 5492B, B&K Precision Inc., USA). All of the measurements were interfaced with a personal computer, which recorded the data automatically. The samples were subjected to a dynamic loading with an average load of 1000 N with amplitude of 1000 N for 12 cycles, as shown in Figure 5. The frequency of the loading was set to be 0·1 Hz. All the samples were tested at room temperature (22°C ± 2°C).

Experimental results

The experimental results on the samples prepared using the two dispersing methods are introduced and presented below.

Figure 4 Laboratory test set-up: (a) full experimental set-up; (b) close-up view of samples under loading

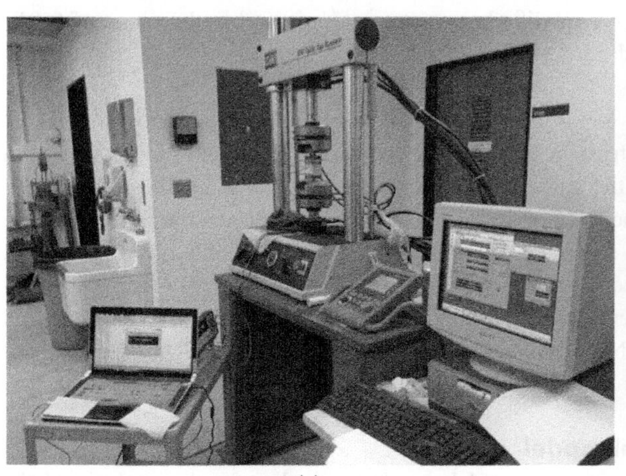
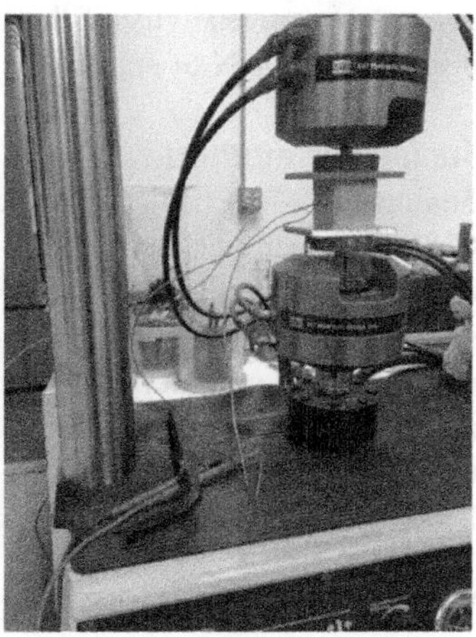

(a) (b)

Figure 5 Dynamic loading curve

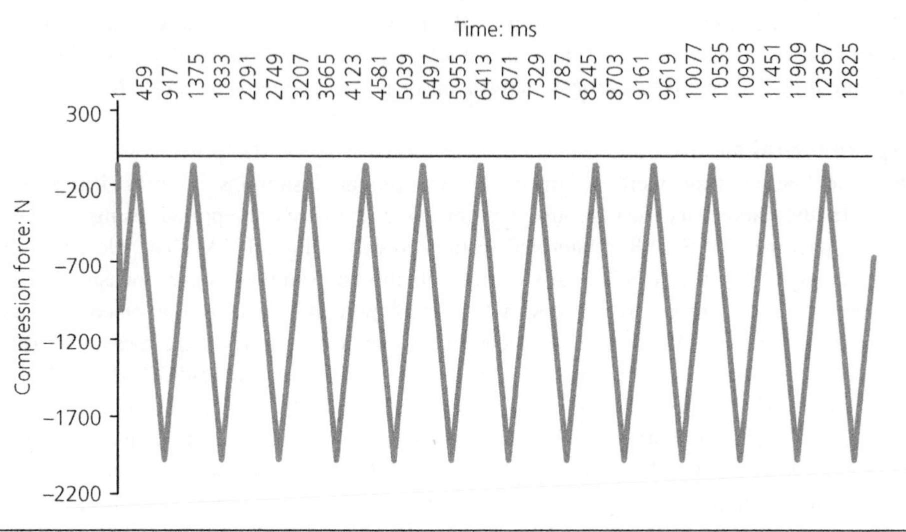

Direct mixing method

Figures 6(a)–6(c) show the piezo-resistive responses of the three cement mortars mixed with CNTs using the direct mixing method under the dynamic loading shown in Figure 5. As can be seen, the electrical resistance changes almost linearly with the compressive stress and the changes are proportional to the stress levels. Figure 6(d) provides the summary for the three samples. The average piezo-resistance, maximum and minimum resistance under the dynamic loading are illustrated in Figure 6(d). It can be seen that, with the direct mixing method, the average piezo-resistivity of the CNT-modified cement mortars is within the range of 0·01–0·06 Ω for 1000 N of applied loading.

Surfactant method

Figures 7(a)–7(c) depict the electrical resistance responses of the cement mortar mix with CNTs that were dispersed using the surfactant method under dynamic loading. The piezo-resistance changes linearly with the compressive stress and the changes are proportional to the stress levels. Figure 7(d) shows a summary of the results, with the average piezo-resistance, maximum and minimum resistance under the dynamic loading illustrated. The average piezo-resistivity of two CNT-modified cement mortar samples using the surfactant method are within the range of 0·2–0·3 Ω for 1000 N of applied loading. One of the samples has a somewhat lower piezo-resistance, within the range of 0·02–0·035 Ω of average piezo-resistivity.

Validation of the theoretical model

To assess the theoretical transfer function model developed earlier for the analysis of the effects of dispersing methods on piezo-resistance in CNT-modified cementitious materials,

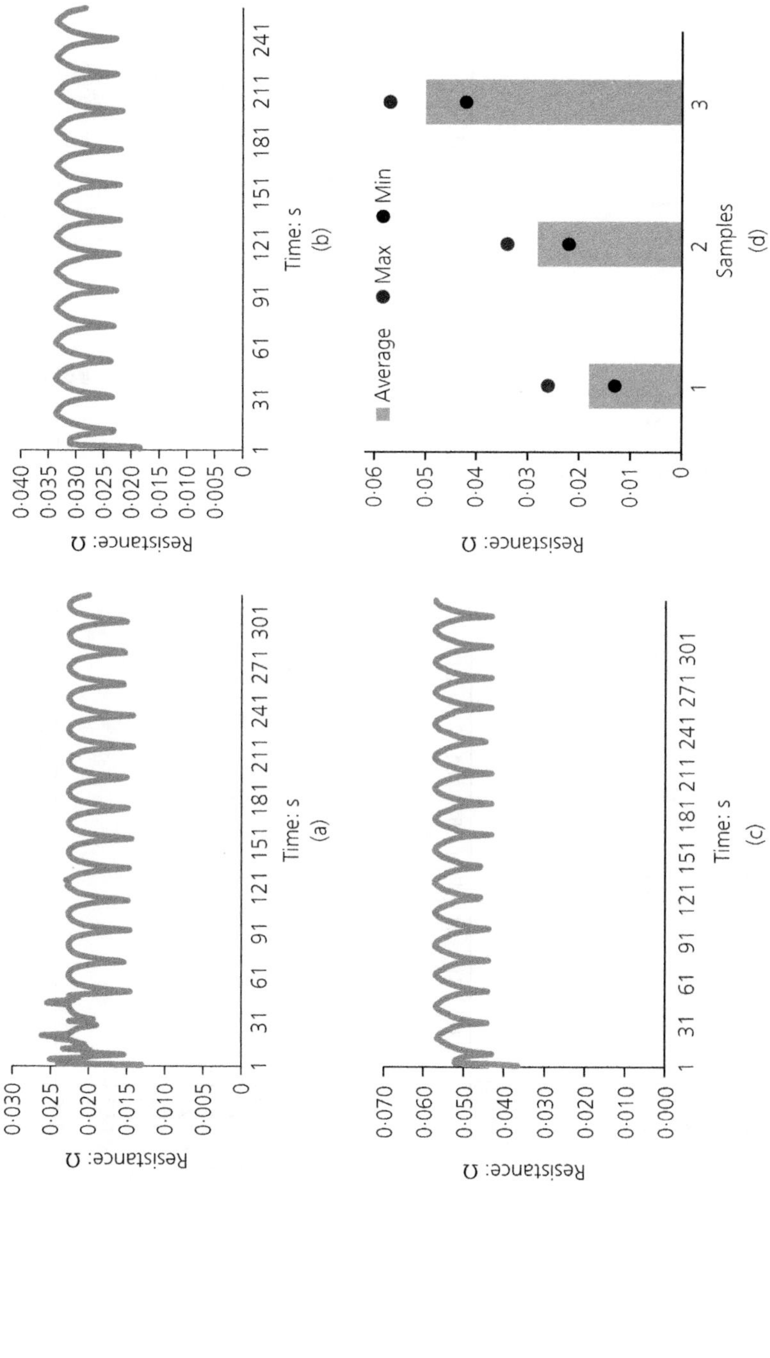

Figure 6 Piezo-resistance responses for direct mixing: (a) sample no. 1; (b) sample no. 2; (c) sample no. 3; (d) summary of three samples

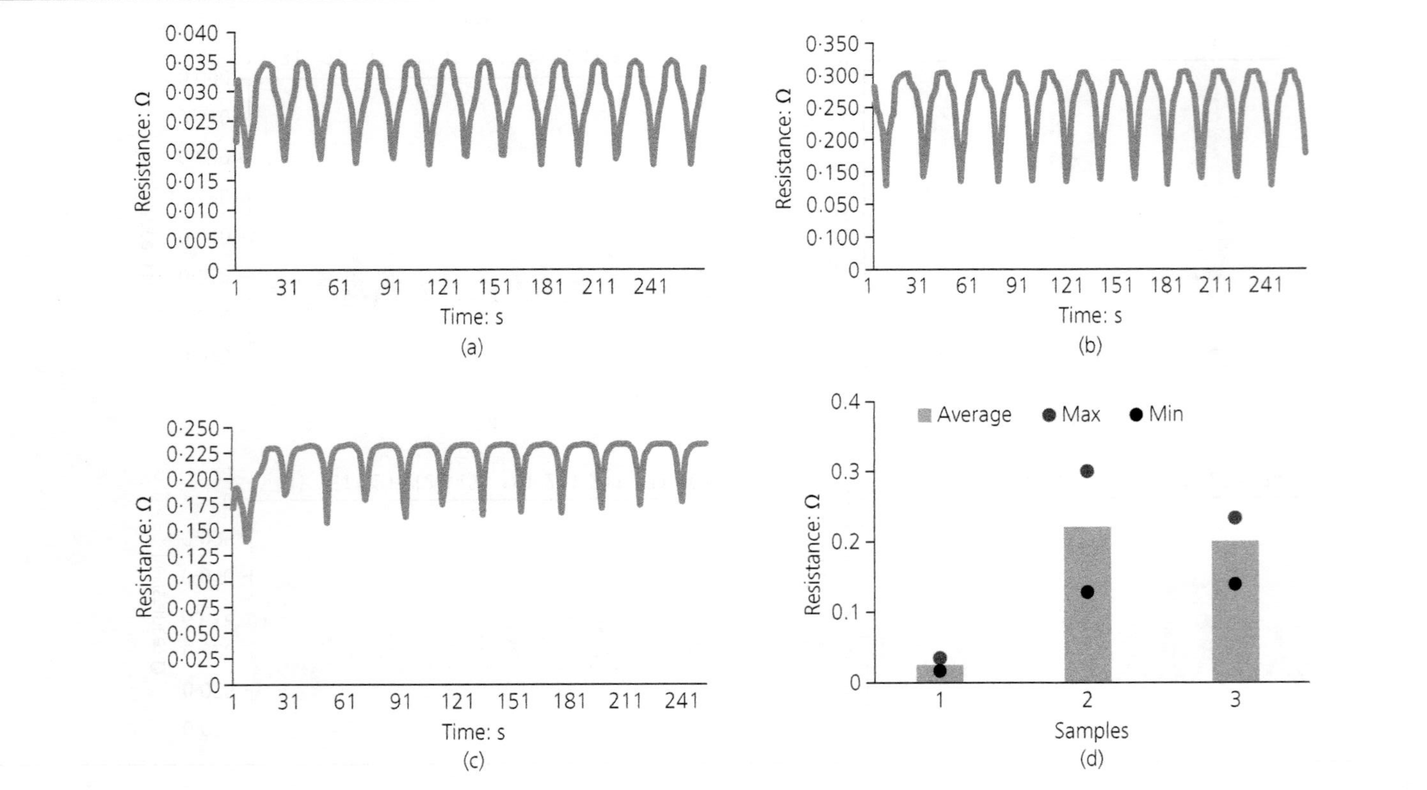

Figure 7 Piezo-resistance responses for surfactant method: (a) sample no. 1; (b) sample no. 2; (c) sample no. 3; (d) summary of three samples

the responses from the samples were analysed statistically to find all the coefficients in Equation 8 for the three different dispersing methods; Equation 8 is repeated below

$$\Delta R = C_A K_B + k_1 \varepsilon - K_B e^{k_2 \varepsilon}$$

The dynamic load varies from 0 to 2000 N with amplitude of 1000 N. For static testing, the amplitude of the dynamic load, 1000 N, was applied and analysed as shown in Figure 8 with static load from 0 to 1000 N. The strains in Equation 8 were calculated using the relationship between load and the surface area of the sample, as below

$$\varepsilon = \frac{P}{EA} \tag{9}$$

Figures 9(a)–9(c) plot the piezo-resistance responses with compression strain changes under the static loading process for the direct mixing method of CNTs. As the piezo-resistance response of the samples prepared using direct mixing is very small and the distribution of the nanotubes in cement matrix is uncertain, the responses do not follow the pattern of an exponential function as in Equation 8, but show a rather linear behaviour; this indicates that the piezo-resistance effects of CNT-modified cementitious materials are mostly contributed by the intrinsic resistance rather than the intertube resistance. For the direct mixing methods, Equation 8 does not fit well for the responses because the dispersing is not effective.

Figures 10(a)–10(c) plot the piezo-resistance responses with compression strain changes under a static loading process for the surfactant method of dispersal for CNTs. The fitted responses using Equation 8 with the surfactant method are also seen in Figures 10(a)–10(c). Figure 10 clearly demonstrate that the piezo-resistances of the samples using surfactant method follow the predictions of response using Equation 8 well. This indicates that the surfactant dispersing method dispersed the CNTs effectively so that the intertube resistance of CNTs also contributes to the piezo-resistance effects and dominates the performance of the piezo-resistive effects instead of intrinsic resistance. This also affirms that Equation 8 can predict the piezo-resistance responses well if the dispersing method effectively disperses the CNTs in the cementitious matrix.

Figure 8 Static loading curve

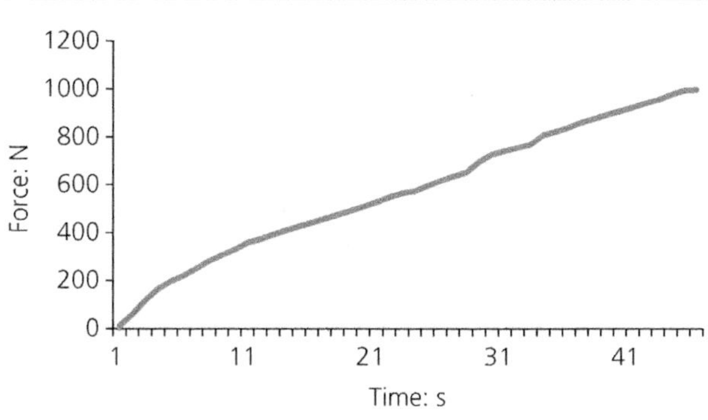

Figure 9 Piezo-resistances plotted against strains of samples with direct mixing methods and their fitted lines using Equation 9 for: (a) sample no. 1; (b) sample no. 2; (c) sample no. 3

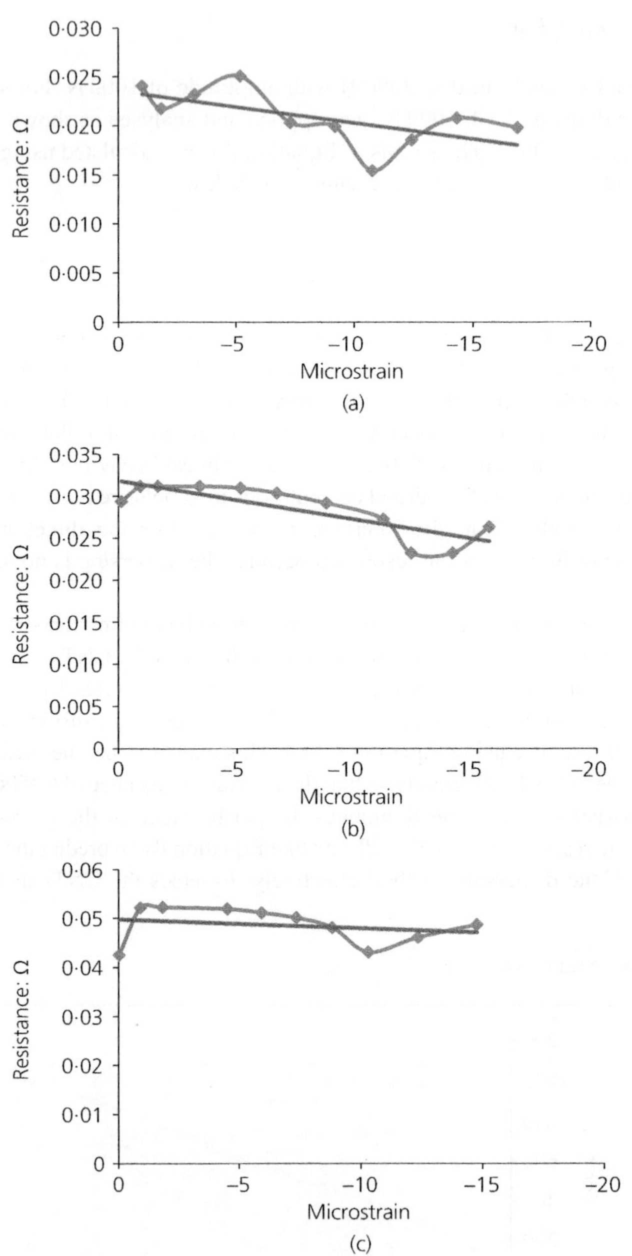

Figure 10 Piezo-resistances plotted against strains of samples with surfactant methods and their fitted lines using the Equation 9 for: (a) sample no. 1; (b) sample no. 2; (c) sample no. 3

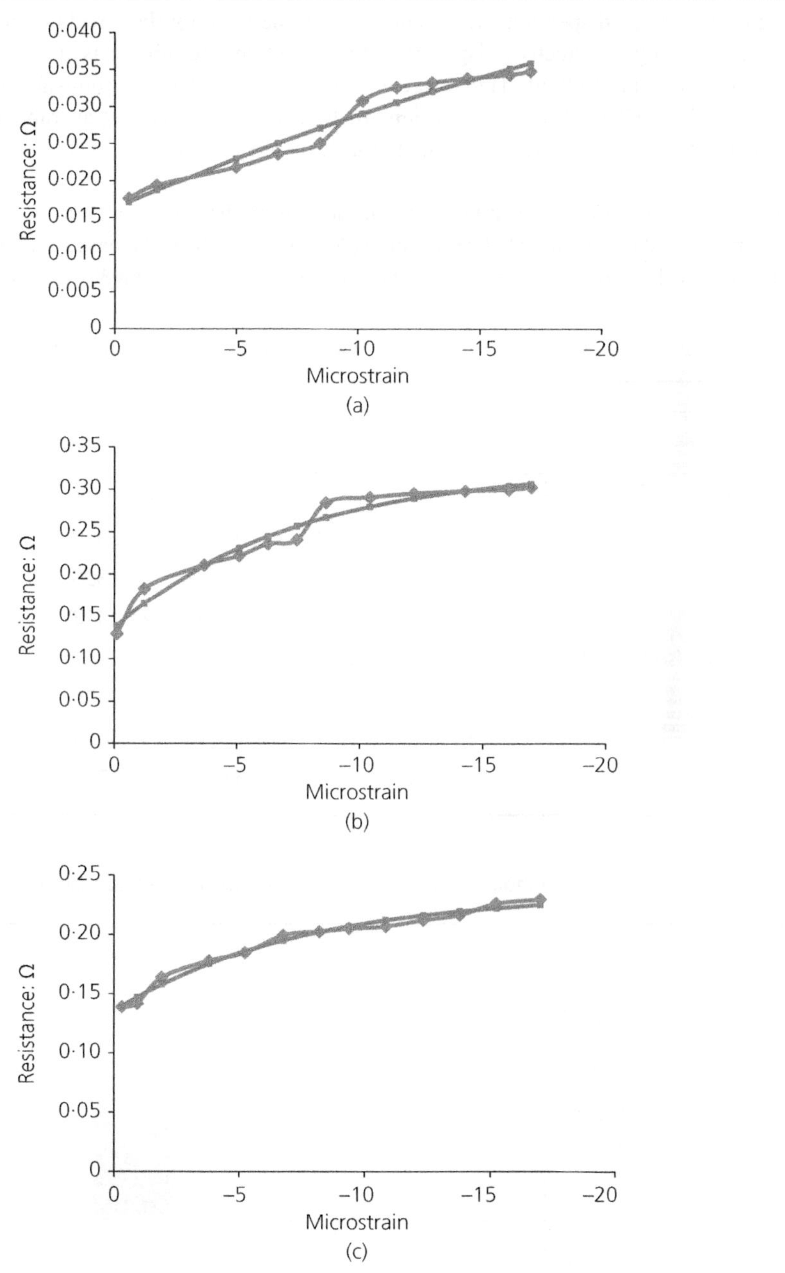

Table 2 summarises all the coefficients of dispersal in Equation 8 for the two methods used to disperse the CNTs. The average K_B and the standard deviation with samples containing CNT dispersed using the different methods are plotted in Figure 11. It can be seen from Table 2 and Figure 11 that the dispersing coefficient, K_B, in Equation 8 for the direct mixing method yields zero, indicating no effective dispersal and no effective intertube resistance is activated using the direct mixing method. The surfactant method showed very promising dispersing effectiveness for CNT-modified cementitious materials, with the largest average dispersing coefficient, K_B, of $0\cdot1271$ and a small standard deviation of $0\cdot09154$.

If the required strain measurement resolution from the smart cement is $10\,\mu\varepsilon$, a positive electrical resistance change to be measured ($\Delta R > 0$) from the CNT-based cementitious material prepared by the surfactant dispersed method, will require a minimum K_B value of $0\cdot14$ according

Table 2 Summary of coefficients in Equation 9 for all groups with three different dispersing methods

Method	Parameters	Sample			Average	Standard deviation
		1	2	3		
Method 1 (direct method)	K_B	0	0	0	0	0
	k_1	0·00009	0·00020	−0·00020	0·00003	0·000207
	k_2	15·5649	17·541	1·31896	11·475	8·85067
	C_A	0·90261	0·74185	2·66418	1·43621	1·06648
Method 2 (surfactant)	K_B	0·05256	0·22927	0·09947	0·1271	0·09154
	k_1	0·00113	0·01143	0·00509	0·00588	0·0052
	k_2	0·02717	0·13573	0·13484	0·09925	0·06242
	C_A	0·30917	0·69794	1·3579	0·78834	0·53018

Figure 11 Average and standard deviation for value K_B for samples containing CNT only

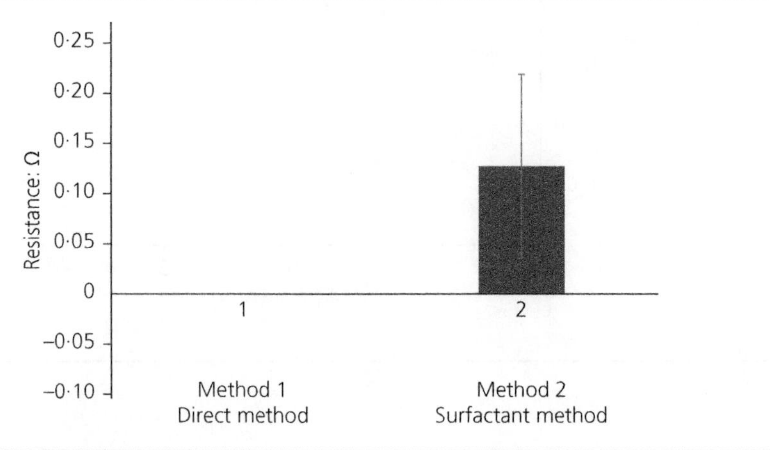

to Equation 8, using the average values calibrated in Table 2 for k_1 (0·00588), k_2 (0·09925) and C_A (0·78834). If the required strain measurement resolution is 15 µε, the minimum K_B value should be 0·105 according to Equation 9. An average dispersing coefficient of 0·1271, as in the validation experiments, will produce a resolution of 12 µε.

In addition, Table 2 shows that the intrinsic piezo-resistances for the surfactant method yield a linear response coefficient from intrinsic resistance, k_1, in a range between 0·001 and 0·006 with small standard deviation, indicating that the intertube resistances in an effectively dispersed matrix will dominate the piezo-resistance effect. It is clearly seen from the figures and table that the theoretical model developed can clearly identify the influence of the dispersing effectiveness from different mixing methods.

Conclusions and future work

This paper has developed a theoretical transfer function model to analyse the dispersing effectiveness of various mixing methods of CNT-modified cementitious materials for the purpose of self-sensing using the piezo-resistance effect. The theoretical transfer function model was tested and validated using two different methods of dispersing CNTs in cementitious materials: the direct mixing method and the surfactant method. The conclusions of this study can be drawn as follows.

(*a*) A theoretical transfer function to analyse the piezo-resistances in CNTs modified cementitious materials has been derived with consideration of dispersing effectiveness.
(*b*) The laboratory tests showed that the developed theoretical model can easily identify the dispersing effectiveness of mixing methods of CNTs in modified cementitious materials.
(*c*) From the experimental results, it has also been shown that the direct method cannot uniformly disperse CNTs in cementitious materials, resulting in small piezo-resistance effects and no activation of intertube resistances of CNTs.
(*d*) The surfactant method is an effective approach to disperse CNTs in cementitious materials, with high dispersing coefficient and small standard derivation.

The transfer function developed in this study can serve as a reference to control the quality of smart cementitious materials, to test the sensing characteristics of these smart materials and to guide the development of new dispersing methods of CNTs in cementitious materials. Future efforts will be devoted to applying the developed theoretical transfer function model in the evaluation of the dispersion effectiveness of further dispersing methods and to develop new dispersing methods, using the guidance of the theoretical model to seek an increase in sensitivity and improved resolution of the piezo-resistance response in CNT-modified smart cementitious materials.

Acknowledgements

This study is partially supported by an NSF grant under ND EPSCoR agreement IIA-1355466 and NDSU Advance Forward program, and a US DOT PHMSA grant under agreement no. DTPH56-15-H-CAAP06. The findings and opinions expressed in this paper are those of the authors only and do not necessarily reflect the views of the sponsors. The authors would like to thank Dr Achintya Bezbaruah, Dr Dinesh Katti and their graduate student for the technical help during this research.

REFERENCES

Ativitavas N, Pothisiri T and Fowler TJ (2006) Identification of fiber-reinforced plastic failure mechanisms from acoustic emission data using neural networks. *Journal of Composite Materials* **40(3)**: 193–226.

Azhari F and Banthia N (2012) Cement-based sensors with carbon fibers and carbon nanotubes for piezoresistive sensing. *Cement and Concrete Composites* **34(7)**: 866–873.

Berger C, Yi Y, Wang Z and De Heer W (2002) Multiwalled carbon nanotubes are ballistic conductors at room temperature. *Applied Physics A* **74(3)**: 363–365.

Dzenis YA and Saunders I (2002) On the possibility of discrimination of mixed mode fatigue fracture mechanisms in adhesive composite joints by advanced acoustic emission analysis. *International Journal of Fracture* **117(4)**: 23–28.

Farrar CR and Worden K (2007) An introduction to structural health monitoring. *Philosophical Transactions of the Royal Society of London: Mathematical, Physical and Engineering Sciences* **365(1851)**: 303–315.

Frank S, Poncharal P, Wang Z and De Heer WA (1998) Carbon nanotube quantum resistors. *Science* **280(5370)**: 1744–1746.

Ganesh E (2013) Single walled and multi walled carbon nanotube structure, synthesis and applications. *International Journal of Innovative Technology and Exploring Engineering (IJITEE)* **2(4)**: 311–320.

Giurgiutiu V, Zagrai A and Bao JJ (2002) Piezoelectric wafer embedded active sensors for aging aircraft structural health monitoring. *Structural Health Monitoring* **1(1)**: 41–61.

Han B and Yu X (2014) Effect of surfactants on pressure-sensitivity of CNT filled cement mortar composites. *Frontiers in Materials* **1**: 27.

Han B, Yu X, and Ou J (2014) *Self-Sensing Concrete in Smart Structures*. Elsevier, Waltham, MA, USA, pp. 47–56.

Heyd R, Charlier A and McRae E (1997) Uniaxial-stress effects on the electronic properties of carbon nanotubes. *Physics Review B* **55(11)**: 6820.

Kim H, Park I and Lee H (2014) Improved piezoresistive sensitivity and stability of CNT/cement mortar composites with low water–binder ratio. *Composite Structures* **116**: 713–719.

Kon S, Oldham K and Horowitz R (2007) Piezoresistive and piezoelectric MEMS strain sensors for vibration detection. *Proceedings of the 14th International Symposium on Smart Structures and Materials and Nondestructive Evaluation and Health Monitoring. San Diego, CA, USA,* pp. 65292V–65292V.

Liang W, Bockrath M, Bozovic D *et al.* (2001) Fabry–Perot interference in a nanotube electron waveguide. *Nature* **411(6838)**: 665–669.

Lyshevski SE (2002) *MEMS and NEMS: Systems, Devices, and Structures*. CRC Press, Boca Raton, FL, USA.

Maiti A, Svizhenko A and Anantram M (2002) Electronic transport through carbon nanotubes: effects of structural deformation and tube chirality. *Physical Review Letters* **88(12)**: 126805.

Materazzzi AL, Ubertini F and D'Alessandro A (2013) Carbon nanotube cement-based sensors for dynamic monitoring of concrete structures. *Proceedings of the 2013 FIB Symposium. Tel Aviv, Israel,* pp. 22–24.

Minot E, Yaish Y, Sazonova V *et al.* (2003) Tuning carbon nanotube band gaps with strain. *Physical Review Letters* **90(15)**: 156401.

Naidu PK, Pulagara NV and Dondapati RS (2014) Carbon nanotubes in engineering applications: a review. *Progress in Nanotechnology and Nanomaterials* **3(4)**: 79–82.

Saito Y (2010) *Carbon Nanotube and Related Field Emitters: Fundamentals and Applications.* John Wiley & Sons, Inc, Hoboken, NJ, USA.

Sharma VK, Jelen F and Trnkova L (2015) Functionalized solid electrodes for electrochemical biosensing of purine nucleobases and their analogues: a review. *Sensors* **15(1)**: 1564–1600.

Thostenson ET and Chou TW (2008) Carbon nanotube-based health monitoring of mechanically fastened composite joints. *Composites Science and Technology* **68(12)**: 2557–2561.

Wang X, Li Q, Xie J *et al.* (2009) Fabrication of ultralong and electrically uniform single-walled carbon nanotubes on clean substrates. *Nano Letters* **9(9)**: 3137–3141.

Yang L and Han J (2000) Electronic structure of deformed carbon nanotubes. *Physical Review Letters* **85(1)**: 154.

Yang L, Anantram M, Han J and Lu J (1999) Band-gap change of carbon nanotubes: effect of small uniaxial and torsional strain. *Physics Review B* **60(19)**: 13874.

Yu X and Kwon E (2009) A carbon nanotube/cement composite with piezoresistive properties. *Smart Materials and Structures* **18(5)**: 55010.

Yu X and Kwon E (2012) *Carbon Nanotube Based Self-Sensing Concrete for Pavement Structural Health Monitoring*. University of Minnesota Duluth, Duluth, MN, USA, Research report, Contract Number: US DOT: DTFH61-10-C-00011.

Zamkov M, Alnaser AS, Shan B, Chang Z and Richard P (2006) Probing the intrinsic conductivity of multiwalled carbon nanotubes. *Applied Physics Letters* **89(9)**: 93111.

Zou Y, Tong L and Steven G (2000) Vibration-based model-dependent damage (delamination) identification and health monitoring for composite structures–a review. *Journal of Sound and Vibration* **230(2)**: 357–378.

Dhir and Paine
ISBN 978-0-7277-6457-7
https://doi.org/10.1680/icetsc.64577.159
ICE Publishing: All rights reserved

Chapter 10

Measurement and modelling of conduction in carbon fibre-cement composites

Jing Xu
Research assistant, Key Laboratory of Advanced Civil Engineering Materials (Tongji University), Ministry of Education, Shanghai, China

Wu Yao
Professor, Key Laboratory of Advanced Civil Engineering Materials (Tongji University), Ministry of Education, Shanghai, China

The present work investigated the conductive performance of carbon fibre-reinforced cement under different frequencies of alternating current, involving fibre fractions both below and above the percolation threshold. The electrical conductivity of carbon fibre-reinforced cement changes according to the frequency of the alternating current. Tests on samples with fibre fractions below the percolation threshold show similar change pattern in the impedance–frequency curve. The trend of decreasing impedance with increasing frequency becomes less marked as the carbon fibre content increases. An equivalent circuit model of carbon fibre-reinforced cement has been proposed to describe the electrical performance based upon the experimental results. Moreover, using linear regression the correlation coefficients were analysed to show the rationality of the model. Results indicate that the resistance under alternating current is lower than that under direct current. The capacitance increases with increasing fibre content.

Introduction

Cement-based composite is attractive and has been used in building and construction for centuries before other materials owing to its on-site fabrication, high compressive strength and excellent durability. Normal cement material, however, is a poor electrical conductor and low in tensile strength, especially under dry conditions. For the purpose of improving the strength and ductility of cement, a variety of fibres are usually added into the cement matrix. Carbon fibre-reinforced cement (CFRC) is one type of composite material which is provided with not only excellent mechanical properties, but also high conductivity because carbon fibre has a low resistivity. Therefore, the conductive behaviour of CFRC is now attracting more and more attention from researchers around the world (Cao and Chung, 2004; Chen *et al.*, 2004; Chiarello and Zinno, 2005; Chung, 2001; Fu and Chung, 1997; Reza *et al.*, 2001, 2004; Sun *et al.*, 2002; Wen and Chung, 2006a; Yao *et al.*, 2003).

Electrical conduction in CFRC involves electrons and ions. In the wet state, ionic conduction is more significant than electronic conduction. The current passes through the matrix mainly by cement paste. In the dry state, electronic conduction dominates and carbon fibres are responsible for this conduction (Wen and Chung, 2006b). This investigation is focused on the CFRC under dry state. Furthermore, sand was introduced as an ingredient in order to enhance the strength and volume stability of CFRC. As the conductivity of carbon fibres is much larger than sand and cement paste in a dry state, the principal contribution to the electrical conduction is expected to be electronic. Generally, there are two basic types of electronic conduction in CFRC – the network structure effect and the tunnelling effect. The former is achieved by the motion of free electrons in a large number of carbon fibres, which are tiny and contact each other, forming a continuous network through the entire matrix eventually. The latter is attained by the penetration of free electrons in the cement matrix between carbon fibres which are close enough. The free electrons which possess energy exceeding the activation energy can hop across the barrier formed by the cement matrix (Li *et al.*, 2006; Likalter, 2001; Wen and Chung, 2001). At the same time, there are many more fibres isolated by the cement matrix that carriers cannot get through, which leads to a lot of tiny capacitors existing in CFRC. These capacitors can be ignored under direct current (DC), but they should be considered as electrical components in CFRC under alternating current (AC). This paper presents a simplified equivalent circuit model of CFRC based on the analysis of the relationship between impedance and frequency, aiming at a detailed description of its electrical performance under AC in the intermediate frequency range.

Experimental methods

Materials and composite fabrication

The carbon fibres were isotropic PAN-based (polyacrylonitrile), unsized, and 5–10 mm long, as obtained from SGL Carbon Co. (SGL, Germany). The fibre properties are shown in Table 1. Ordinary Portland cement which corresponded to ASTM Type I was used throughout. The sand used was local natural sand with specific gravity of 2·65. Silica fume (Elkem Materials) was used along with fibres in the amount of 15% by mass of cement. The defoamer, which was used in the amount of 0·13 vol.%, was tri-butyl-phosphate. The water-reducing agent (Kao Chemical Co., 21s; polycarboxylic type water-reducer) was used in the amount of 1·0% by mass of cement. The water/cement ratio was 0·45 (by mass) and the sand/cement ratio was 1·0 (by mass).

CFRC specimens containing 0·05, 0·10, 0·20, 0·30 and 0·40% by volume of fibres with respect to the total volume of the composite were prepared in a laboratory mortar mixer. A rotary mixer with a flat beater was used for mixing. Fibres were added in water and then the

Table 1 Properties of carbon fibre

Diameter: μm	Density: g/cm^3	Tensile strength: GPa	Young's modulus: GPa	Elongation at break: %	Carbon content: $w_c\%$	Electrical resistivity: $\mu\Omega\,m$
$7\pm0\cdot2$	1·78	>3·0	220~240	1·25~1·60	>95	15

defoamer was added and mixed for about 1 min. Then the mixture, cement, sand, silica fume and water-reducing agent were mixed in the mixer for 3 min. After pouring into moulds, an external vibrator was used to facilitate compaction and decrease the amount of air bubbles. The specimens were demoulded after 24 h and then cured in a moist-curing room for 27 days. Eventually all specimens were oven-dried, heating to 115°C for several hours to constant weight.

Testing

In any case, for a specimen under AC, there exists

$$Z = \frac{V}{A} \tag{1}$$

where Z is the impedance of CFRC in Ohm (Ω), V is the measured voltage, and A is the measured current (Cao and Chung, 2004). The impedance Z is related to the resistance R and the capacitance C. During the AC impedance measurement, the resistance of the specimen always exists, whose value equals the testing result under DC. So that DC measurements were conducted along with AC impedance measurements, where the measured resistance was assumed equal to the impedance of the specimen when the frequency was 0 Hz, both the AC and the DC measurements were conducted using the four-probe method in which the electrical contact and polarisation effect are negligible. Two Fluke 89 multimeters were used.

The specimen configuration for testing is illustrated in Figure 1. The specimen was a rectangular bar of size 40 mm × 40 mm × 160 mm. It consisted of a silver coating on both of the 40 mm × 40 mm sides. Electrical contacts in the form of silver coating in conjunction with copper wire were applied on the side surface and around the entire 40 mm × 40 mm perimeter of the bar. Three specimens of each composition were tested.

Figure 1 Specimen configuration, all dimensions are in millimetres

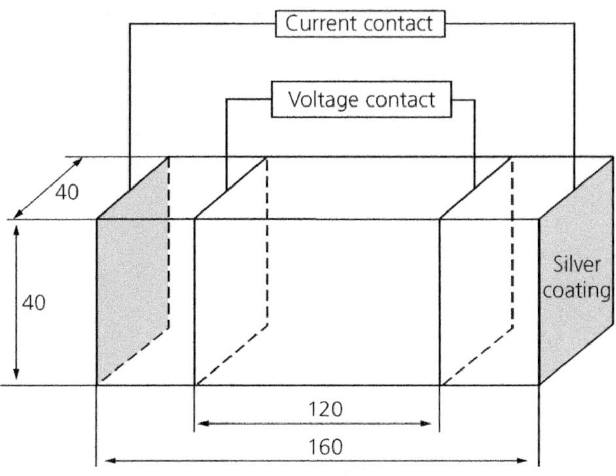

In the present study, AC impedance measurements were conducted by a potentiostat/ galvanostat (Type M273A) connected with a lock-in amplifier (Type M5210). All were manufactured by EG&G PARC (Princeton Applied Research Co., USA). Measurement and data processing were supported by software EIS-M398 provided by the same manufacturer. The amplitude of the sinusoidal voltage was chosen to be 5 mV. Frequency was swept from 10 Hz to 100 kHz (five points per decade). As the four-probe method was used, the current and voltage were recorded simultaneously while the frequency was sweeping, and then converted into impedance by numerical program.

Results and discussion

Figure 2 shows the DC conductivity values plotted against volumetric fraction of carbon fibres for specimens at 28-day hydration. It can be seen that a narrow zone, which is defined as the percolation transition zone, was between the fibre content of 0·1 and 0·3 vol.%. Here, the percolation threshold was around 0·2 vol.% (Cao and Chung, 2004; Wen and Chung, 2001). Testing results of the AC resistivity are listed in Table 2. It can be seen that as the frequency of the applied current was increased, the value of the impedance was reduced.

All the impedance values were normalised with respect to their initial values, that is, the reading at 10 Hz, as displayed in Figure 3. It shows that the descending trend of impedance along with frequency became less marked when the content of carbon fibre increased. The probability of adjacent fibres contacting with each other rises as the fibre content increases. Thus at relative high fibre content, CFRC approaches a pure resistance. Figure 3 also shows that the impedance– frequency curves follow a similar pattern. For the cases of fibre fractions below or around the percolation threshold, the impedance decreases with increasing frequency, such that the decrease is more significant between frequencies of 100 Hz and 1 kHz, owing to the high contribution of capacitive reactance. For the cases of fibre fractions above the percolation threshold, the steepest region is above 1 kHz and the curves are much closer to being a straight line, owing to a remarkable reduction in the capacitive reactance proportion in CFRC. Therefore, the capacitance has a low impact on the conductive system when the frequency varies.

Figure 2 DC conductivity plotted against carbon fibre content for specimens

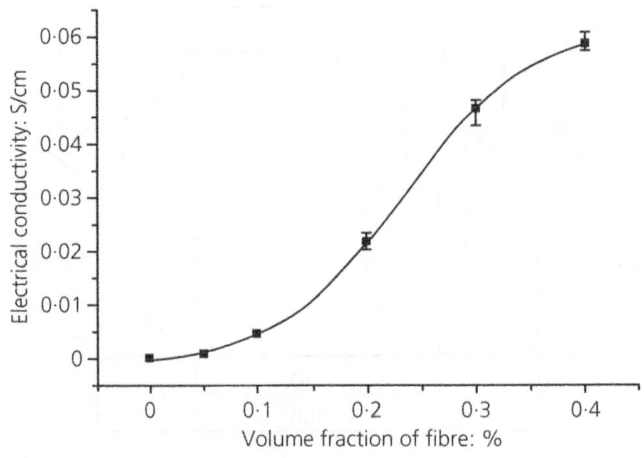

Table 2 The AC resistivity (Ω cm) for different fibre content under varied frequency

Frequency: Hz	Carbon fibre content: vol.%				
	0·05	0·10	0·20	0·30	0·40
10·00	$1{\cdot}34 \times 10^3$	$2{\cdot}30 \times 10^2$	$4{\cdot}15 \times 10^1$	$1{\cdot}98 \times 10^1$	$1{\cdot}66 \times 10^1$
15·85	$1{\cdot}32 \times 10^3$	$2{\cdot}26 \times 10^2$	$4{\cdot}14 \times 10^1$	$1{\cdot}99 \times 10^1$	$1{\cdot}66 \times 10^1$
25·12	$1{\cdot}30 \times 10^3$	$2{\cdot}23 \times 10^2$	$4{\cdot}14 \times 10^1$	$1{\cdot}98 \times 10^1$	$1{\cdot}66 \times 10^1$
39·81	$1{\cdot}26 \times 10^3$	$2{\cdot}16 \times 10^2$	$4{\cdot}11 \times 10^1$	$1{\cdot}98 \times 10^1$	$1{\cdot}66 \times 10^1$
63·10	$1{\cdot}21 \times 10^3$	$2{\cdot}07 \times 10^2$	$4{\cdot}08 \times 10^1$	$1{\cdot}97 \times 10^1$	$1{\cdot}65 \times 10^1$
100·00	$1{\cdot}15 \times 10^3$	$2{\cdot}00 \times 10^2$	$4{\cdot}02 \times 10^1$	$1{\cdot}97 \times 10^1$	$1{\cdot}65 \times 10^1$
158·50	$1{\cdot}10 \times 10^3$	$1{\cdot}91 \times 10^2$	$3{\cdot}95 \times 10^1$	$1{\cdot}97 \times 10^1$	$1{\cdot}64 \times 10^1$
251·20	$1{\cdot}05 \times 10^3$	$1{\cdot}85 \times 10^2$	$3{\cdot}88 \times 10^1$	$1{\cdot}96 \times 10^1$	$1{\cdot}64 \times 10^1$
398·10	$1{\cdot}00 \times 10^3$	$1{\cdot}79 \times 10^2$	$3{\cdot}82 \times 10^1$	$1{\cdot}95 \times 10^1$	$1{\cdot}64 \times 10^1$
631·00	$9{\cdot}68 \times 10^2$	$1{\cdot}73 \times 10^2$	$3{\cdot}75 \times 10^1$	$1{\cdot}95 \times 10^1$	$1{\cdot}63 \times 10^1$
1000·00	$9{\cdot}39 \times 10^2$	$1{\cdot}70 \times 10^2$	$3{\cdot}70 \times 10^1$	$1{\cdot}94 \times 10^1$	$1{\cdot}63 \times 10^1$
1585·00	$9{\cdot}14 \times 10^2$	$1{\cdot}64 \times 10^2$	$3{\cdot}64 \times 10^1$	$1{\cdot}93 \times 10^1$	$1{\cdot}62 \times 10^1$
2512·00	$8{\cdot}91 \times 10^2$	$1{\cdot}61 \times 10^2$	$3{\cdot}59 \times 10^1$	$1{\cdot}91 \times 10^1$	$1{\cdot}61 \times 10^1$
3981·00	$8{\cdot}68 \times 10^2$	$1{\cdot}57 \times 10^2$	$3{\cdot}53 \times 10^1$	$1{\cdot}89 \times 10^1$	$1{\cdot}60 \times 10^1$
6310·00	$8{\cdot}51 \times 10^2$	$1{\cdot}56 \times 10^2$	$3{\cdot}50 \times 10^1$	$1{\cdot}88 \times 10^1$	$1{\cdot}59 \times 10^1$
10 000·00	$8{\cdot}41 \times 10^2$	$1{\cdot}55 \times 10^2$	$3{\cdot}49 \times 10^1$	$1{\cdot}87 \times 10^1$	$1{\cdot}58 \times 10^1$
15 850·00	$8{\cdot}25 \times 10^2$	$1{\cdot}53 \times 10^2$	$3{\cdot}46 \times 10^1$	$1{\cdot}86 \times 10^1$	$1{\cdot}58 \times 10^1$
25 120·00	$8{\cdot}12 \times 10^2$	$1{\cdot}53 \times 10^2$	$3{\cdot}46 \times 10^1$	$1{\cdot}85 \times 10^1$	$1{\cdot}57 \times 10^1$
39 810·00	$8{\cdot}05 \times 10^2$	$1{\cdot}52 \times 10^2$	$3{\cdot}46 \times 10^1$	$1{\cdot}84 \times 10^1$	$1{\cdot}57 \times 10^1$
63 100·00	$7{\cdot}97 \times 10^2$	$1{\cdot}51 \times 10^2$	$3{\cdot}43 \times 10^1$	$1{\cdot}84 \times 10^1$	$1{\cdot}56 \times 10^1$
100 000·00	$7{\cdot}82 \times 10^2$	$1{\cdot}50 \times 10^2$	$3{\cdot}42 \times 10^1$	$1{\cdot}83 \times 10^1$	$1{\cdot}56 \times 10^1$

Figure 3 Normalisation curves of impedance plotted against frequency for different carbon fibre fractions

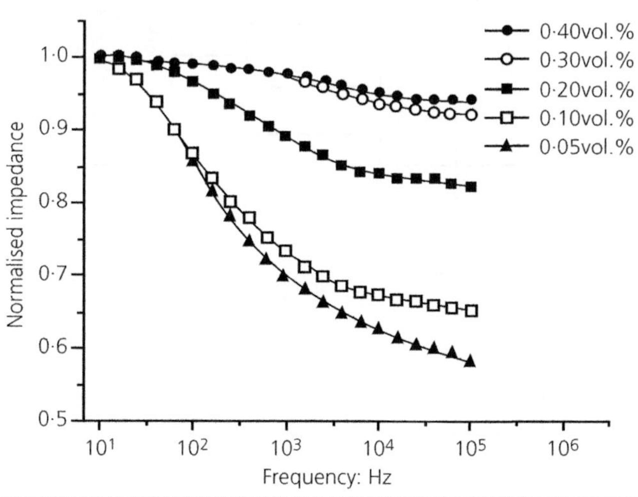

The sketch of fibre distribution in CFRC under dry state is illustrated in Figure 4. Basically, there are three kinds of paths for passing of AC in such a structure: the continuous conductive path (R_c), the tunnelling conductive path (R_t) and capacitive conductive path (C). Each of the paths corresponds to an equivalent circuit element. The continuous conductive path is formed by adjacent fibres making contact with each other. The tunnelling conductive path is attributed to tunnel transmission conduction of electrons from fibre to fibre. Fibres separated by cement form the capacitive conductive path. The contribution of tunnelling conduction and capacitive conduction may significantly decrease due to the increasing fibre content. Above the percolation threshold, continuous network conduction dominates.

First, the most simplified model of the circuit would be all equivalent elements connected in parallel, as illustrated in Figure 5(a). In this case, if the frequency $f \rightarrow +\infty$, then the total

Figure 4 The sketch of carbon fibre distribution in CFRC

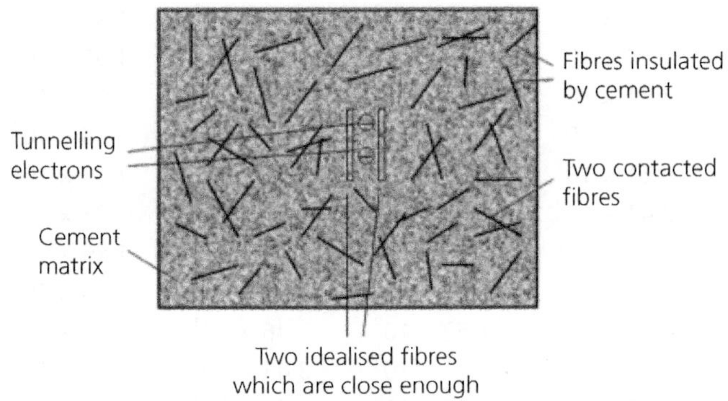

Figure 5 Simplified model of circuit: (a) connected in parallel; (b) connected in series

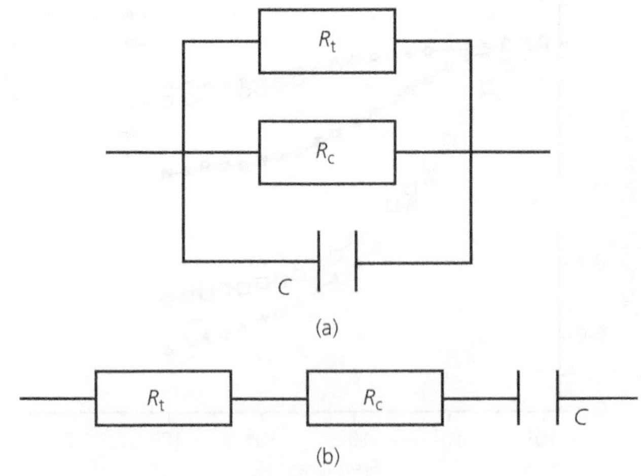

impedance $Z \to 0$. Another circuit model would be equivalent elements connected in series, as shown in Figure 5(b). But if the frequency $f \to 0$, then the impedance $Z \to +\infty$. Therefore, these two simplified models are inconsistent with the actual condition.

An improved equivalent circuit can be taken into account and it is depicted in Figure 6. In this model, R_1 and R_2 are both the combination of R_t and R_c; however, the proportion of R_t and R_c in R_1 and R_2 is different. The relationship among resistance R_1 and R_2, capacitance C and the total impedance Z can be demonstrated using the following equation

$$Z = \frac{R_1}{\sqrt{1 + \omega^2 C^2 R_1^2}} + R_2 \tag{2}$$

where $\omega = 2\pi f$, with f being the applied frequency in Hertz (Hz); C is the capacitance in Farads (F). In the case of DC applied, CFRC can be described as a pure resistance for $f = 0$, and then $Z = R_1 + R_2$. In this equation, R_2 is an unknown parameter, but we can assume that if the AC frequency is high enough, then $R_2 \approx Z$, for the capacitance path can be regarded as short circuit. During the test, when $f = 100$ kHz, R_2 can be obtained from Z, or it can be written as $R_2 = Z|_{f=100 \text{ kHz}}$.

Equation 2 can be rewritten as

$$\frac{1}{(Z - R_2)^2} = \frac{1}{R_1^2} + \omega^2 C^2 \tag{3}$$

Obviously, the relation between $1/(Z - R_2)^2$ and ω^2 is linear, and the value of R_1 and C can be calculated by one-dimensional linear regression. Also, the correlation coefficient r, which can be determined by Equation 4, was calculated in order to verify the degree of correlation.

$$r = \frac{\sum (X - \overline{X})(Y - \overline{Y})}{n\sigma_X \sigma_Y} \tag{4}$$

where X represents ω^2 and Y represents $1/(Z - R_2)^2$; $\overline{X}$ and $\overline{Y}$ are the average values of X and Y; σ_X and σ_Y are the standard deviation of X and Y. Table 3 presents all testing and calculated results of samples with different fibre fractions.

Figure 6 An improved simplified model of circuit in CFRC

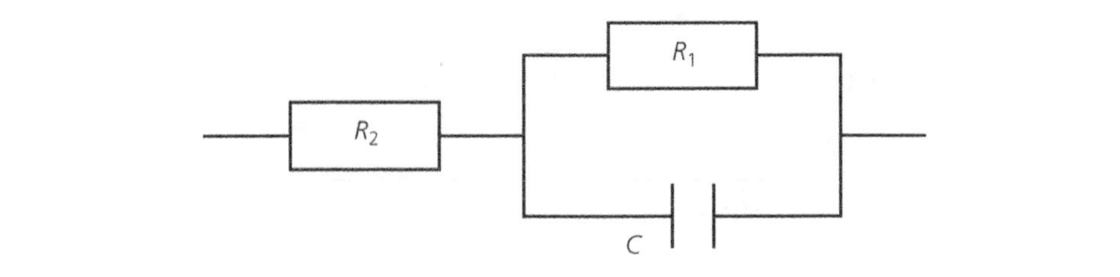

Table 3 Comparison between theoretical and experimental results

Carbon fibre content: vol.%	0·05	0·10	0·20	0·30	0·40
R_1: Ω	$1{\cdot}20 \times 10^2$	$1{\cdot}40 \times 10^1$	$1{\cdot}92 \times 10^0$	$8{\cdot}61 \times 10^{-1}$	$4{\cdot}26 \times 10^{-1}$
R_2: Ω	$7{\cdot}82 \times 10^2$	$1{\cdot}50 \times 10^2$	$3{\cdot}42 \times 10^1$	$1{\cdot}83 \times 10^1$	$1{\cdot}57 \times 10^1$
C: μF	$1{\cdot}75 \times 10^{-1}$	$2{\cdot}54 \times 10^0$	$8{\cdot}60 \times 10^0$	$3{\cdot}01 \times 10^1$	$5{\cdot}48 \times 10^1$
Calculated $R_1 + R_2$ under AC: Ω	$9{\cdot}02 \times 10^2$	$1{\cdot}64 \times 10^2$	$3{\cdot}61 \times 10^1$	$1{\cdot}92 \times 10^1$	$1{\cdot}61 \times 10^1$
Tested $R_1 + R_2$ under DC: Ω	$1{\cdot}66 \times 10^3$	$2{\cdot}36 \times 10^2$	$4{\cdot}61 \times 10^1$	$2{\cdot}16 \times 10^1$	$1{\cdot}71 \times 10^1$
r: %	99·6	98·0	97·0	99·8	98·0

From Table 3, it was found that the corresponding correlation coefficients of samples with fibre fractions below and above percolation threshold were close to 100%, which indicates the rationality of the proposed model has been justified. The resistance, whether calculated or tested, all decreased as the fibre content increased. The total resistances $R_1 + R_2$ under AC are all smaller than the resistance tested under DC. The reason is that the time is too short to form a polarisation layer at the position of both electrodes in the form of silver coating due to the sinusoidal characteristic of AC; thus the electromotive force of polarisation between the two electrodes is absent (Hou *et al.*, 2002). Figure 7 shows the relationship between the proportion of R_1 in the total resistance (both calculated and tested) and fibre content. From the figure, it can be seen that the proportion of R_1 in the total resistance decreased sharply with increasing fibre content and that this decreasing trend became less marked in the case of high fibre content. This phenomenon is attributed to the influence of the tunnelling conduction R_t being greatly weakened by increasing fibre content. In the meantime, we can also deduce that the ratio of R_t/R_1 is higher than that of R_c/R_1. At the same time, from the calculated

Figure 7 The proportion of R_1 in total resistance (calculated and tested) plotted against fibre content

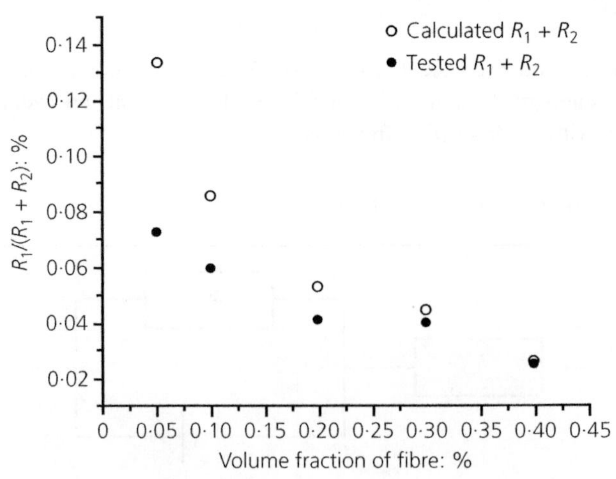

results, the capacitance increases with increasing fibre content for all types of samples. This is attributed to the distance between fibres associated with fibre content. Isolated carbon fibres exist in the CFRC conductive system both below and above the percolation threshold. Above the percolation threshold, a relatively high fibre content results in much closer fibre distance in comparison with that below the percolation threshold. The discontinuous fibres can be treated as a parallel plate capacitor with cement matrix as its dielectric. The total capacitance C would be

$$C = \frac{\varepsilon S}{d} \tag{5}$$

where ε is the permittivity and d is the equivalent distance between fibres (Song, 2000). As d is reduced along with increasing fibre content, C would become larger.

Conclusions

The descending trend of impedance along with increasing frequency becomes less marked when the amount of carbon fibre in CFRC increases under AC. A similar manner of change appears on the impedance–frequency curve of samples with fibre fractions below the percolation threshold and the impedance decreases more significantly in the frequency range of 100 Hz to 1 kHz. Based upon the analysis of AC impedance measurement, an equivalent circuit model, illustrated as a resistance in parallel with a capacitance and then connecting to another resistance, was proposed to describe the electrical performance of CFRC. Testing results were analysed with linear regression and the regression correlation coefficient testified to the rationality of this model. Regression results also indicated that the calculated resistance is smaller than that tested under DC, and the calculated capacitance increases with increasing fibre content.

Acknowledgements

This work was supported by National Natural Science Foundation of China (No. 50238040) and Program for Young Excellent Talents in Tongji University.

REFERENCES

Cao J and Chung DDL (2004) Electric polarization and depolarization in cement-based materials, studied by apparent electrical resistance measurement. *Cement and Concrete Research* **34(3)**: 481–485.

Chen B, Wu K and Yao W (2004) Conductivity of carbon fiber reinforced cement-based composites. *Cement and Concrete Composites* **26(4)**: 291–297.

Chiarello M and Zinno R (2005) Electrical conductivity of self-monitoring CFRC. *Cement and Concrete Composites* **27(4)**: 463–469.

Chung DDL (2001) Electromagnetic interference shielding effectiveness of carbon materials. *Carbon* **39(2)**: 279–285.

Fu X and Chung DDL (1997) Self-monitoring in carbon fiber reinforced mortar by reactance measurement. *Cement and Concrete Research* **27(6)**: 845–852.

Hou Z, Li Z and Tang Z (2002) Contrasting research on AC electrical properties of carbon fiber electrically conductive concrete. *Concrete* **150(4)**: 32–52.

Li H, Xiao H and Ou J (2006) Effect of compressive strain on electrical resistivity of carbon black-filled cement-based composites. *Cement and Concrete Composites* **28(9)**: 824–828.

Likalter AA (2001) Hopping conductivity in granular metals near the insulator–metal transition. *Physica A* **291(1–4)**: 144–158.

Reza F, Batson GB, Yamamuro JA and Lee JS (2001) Volume electrical resistivity of carbon fiber cement composites. *ACI Materials Journal* **98(1)**: 25–35.

Reza F, Yamamuro JA and Batson GB (2004) Electrical resistance change in compact tension specimens of carbon fiber cement composites. *Cement and Concrete Composites* **26(7)**: 873–881.

Song G (2000) Equivalent circuit model for AC electrochemical impedance spectroscopy of concrete. *Cement and Concrete Research* **30(11)**: 1723–1730.

Sun M, Li Z and Liu Q (2002) The electromechanical effect of carbon fiber reinforced cement. *Carbon* **40(12)**: 2273–2275.

Wen S and Chung DDL (2001) Effect of carbon fiber grade on the electrical behavior of carbon fiber reinforced cement. *Carbon* **39(3)**: 369–373.

Wen S and Chung DDL (2006a) Self-sensing of flexural damage and strain in carbon fiber reinforced cement and effect of embedded steel reinforcing bars. *Carbon* **44(8)**: 1496–1502.

Wen S and Chung DDL (2006b) The role of electronic and ionic conduction in the electrical conductivity of carbon fiber reinforced cement. *Carbon* **44(11)**: 2130–2138.

Yao W, Chen B and Wu K (2003) Smart behavior of carbon fiber reinforced cement-based composite. *Journal of Materials Science and Technology* **19(3)**: 239–243.

Sensor monitoring

Dhir and Paine
ISBN 978-0-7277-6457-7
https://doi.org/10.1680/icetsc.64577.171

Chapter 11

A galvanic sensor for monitoring corrosion of steel in carbonated concrete

N. R. Short
Department of Civil Engineering, Aston University, Aston Triangle, Birmingham B4 7ET, UK

C. L. Page
Department of Civil Engineering, Aston University, Aston Triangle, Birmingham B4 7ET, UK

G. K. Glass
Taywood Engineering Ltd, 345 Ruislip Road, Southall, Middlesex UB1 2QX, UK

The construction and operation are described of a galvanic sensor which responds to changes in the corrosion intensity of mild steel in carbonated mortars or concretes. The sensor is simple to construct and operate; it is robust and has potential for development as an inexpensive method of long-term continuous nondestructive corrosion monitoring of steel in concrete. Its possible applications include laboratory and field studies of effects of environmental exposure conditions and cement matrix variables on corrosion rates of steel in reinforced concrete structures. Information derived from such investigations is needed to provide an improved basis for service-life prediction.

Introduction

Carbonation-induced corrosion of reinforcing steel is one of the major causes of the premature deterioration of structural concrete. While factors that affect the rate of carbonation of concrete are well documented,[1,2] this is not true of those that influence the rate of corrosion of steel which has become depassivated as a consequence of carbonation. Variables such as temperature, relative humidity and the presence of chloride salts (even at concentrations that would constitute a low corrosion risk in non-carbonated concrete) are known to influence corrosion rates in carbonated concrete.[3,4] However, their effects have usually been investigated by somewhat elaborate techniques, involving sophisticated electrochemical monitoring procedures of the kinds discussed in recent reviews.[5,6] These tests have also required the use of specimens exposed to accelerated carbonation in atmospheres of enhanced carbon ioxide concentration. While such methods are often the only practical way of rapidly assessing new products, such as cementitious repair materials, it is not known whether or not they simulate realistically the effects of prolonged normal atmospheric exposure conditions.

Although estimates of long-term integrated corrosion losses can be obtained using devices such as electrical resistance probes,[6,7] these are incapable of providing information about short-term changes in corrosion rate in response to changing environmental factors etc. A simple sensor that could monitor the instantaneous corrosion rate of steel continuously over a long period of time would therefore be useful as a means of determining the effect of changes in environmental conditions and cement matrix variables, both in the laboratory and for field studies. Successful monitoring should enable determination of the initiation time to depassivation of steel, the subsequent rate of corrosion as a function of time, and the tolerable amount of corrosion before cracking, for various environmental conditions and concrete qualities. This information would provide an improved basis for service life prediction.[8] Devices could possibly be based on a technique involving measurement of the galvanic current flowing between mild steel anodes and stainless steel (or other) cathodes forming adjacent sections of composite bars. Such sensors have been evaluated for use in atmospheric corrosion monitoring,[9] and more recently to assess relative corrosion rates in concrete.[10–12]

The aim of the work described in this Paper was to design and evaluate a galvanic sensor for monitoring electrochemical changes associated with the initiation and propagation of carbonation-induced corrosion. In particular, the sensor was required to be sufficiently sensitive to detect changes in corrosiveness of the concrete as a result of, for example, changes in moisture conditions and the presence of chlorides. It was also necessary that the sensor should be easy to construct, and sufficiently robust for use in field studies, and that the associated instrumentation required for measurement should be simple.

Experimental details

The galvanic sensor consisted of a disc of mild steel (anode) exposed through a hole drilled perpendicular to the axis of an austenitic (AISI type 304) stainless steel bar (cathode), as shown in Fig. 1. To insulate the mild steel from the stainless steel and to prevent crevice attack,

Fig. 1 Construction of the galvanic sensor, showing detail of mild steel/stainless steel interface

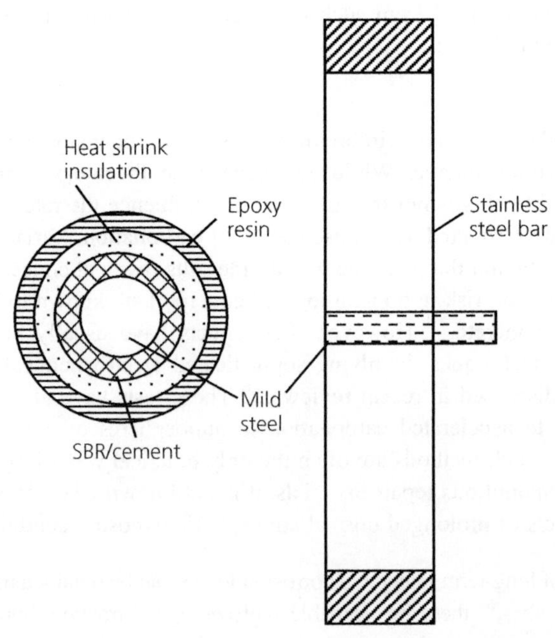

the former was covered by a thin layer of styrene-butadiene rubber (SBR) modified cement paste, and then by heat shrink tubing, and was fixed within the stainless steel bar by an epoxy adhesive. A capillary salt bridge was mounted so that its tip was in close proximity to the exposed mild steel.

The sensor's primary mode of operation was as a galvanic couple, in which the current (I_{galv}) provided an index of the corrosiveness of the surrounding mortar. In preliminary experiments the anode/cathode area ratio was altered by varying the diameters of the mild steel and stainless steel bars. This work indicated that suitably high galvanic activity was achieved at anode/cathode area ratios of 1 : 200, the exposed area of mild steel being approximately 0.2 cm^2. The device could also be used to form a cell subjected to polarization by means of a potentiostat with the mild steel disc as working electrode and the stainless steel bar as counter electrode. This allowed the determination of polarization resistance R_p, uncompensated electrolyte resistance R_{unc} and corrosion potential E_{corr}.

To ascertain whether or not any artefacts were associated with measurements made with the galvanic sensor, an alternative polarization resistance probe was designed, consisting of a mild steel sheet of exposed area 2 cm^2 fixed parallel to a mesh counter electrode (titanium wire coated with an electro-catalyst) with a capillary salt bridge terminating in the space between them (Fig. 2). The cell arrangement promoted a more uniform current distribution than the galvanic sensor, and the working electrode area was increased to facilitate accurate current measurement.

Fig. 2 Construction of the polarization resistance sensor

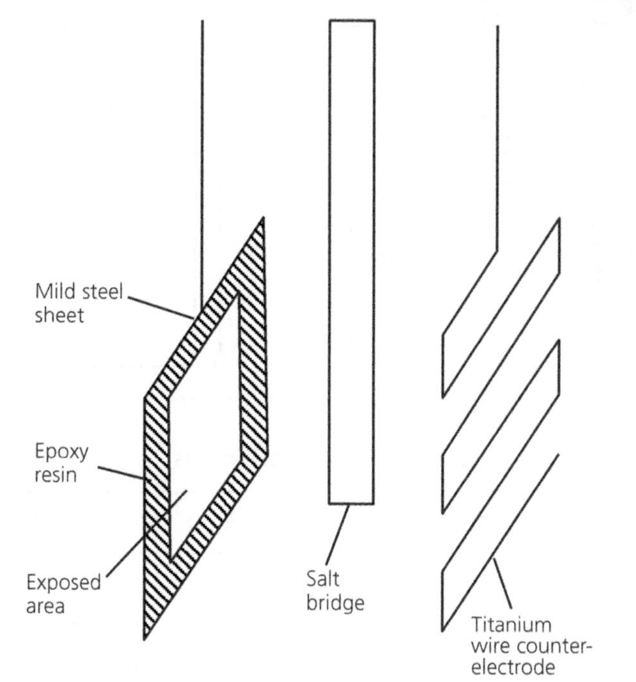

Both types of sensor were then embedded in mortar cubes as shown in Fig. 3. The cements used included an ordinary Portland cement (OPC) and blends prepared from OPC and 65% ground granulated blastfurnace slag (ggbs) or 30% pulverized fuel ash (PFA). The compositions of the various materials, expressed as percentages by weight of the constituent oxides, are given in Table 1. A medium grade sand, an aggregate/cement ratio of 6:1 and a high water/cement ratio (0·9) were used; samples were cured for 24 h before being demoulded and exposed to laboratory air. These mixes were chosen so that the mortars would have a relatively open pore structure to allow rapid carbonation and a fast response to changes in relative humidity (RH). Some specimens contained chloride, which was introduced by dissolving analytical reagent grade sodium chloride in the water before mixing, giving a chloride ion concentration of either 0.4 or 1 .0% by weight of the cement.

Environmental chambers of the form shown in Fig. 4 were constructed for carrying out accelerated carbonation at controlled temperature/relative humidity, and subsequently to create atmospheres of systematically varied RH. In a preliminary experiment, it was

Fig. 3 Location of the sensors in the mortar cubes

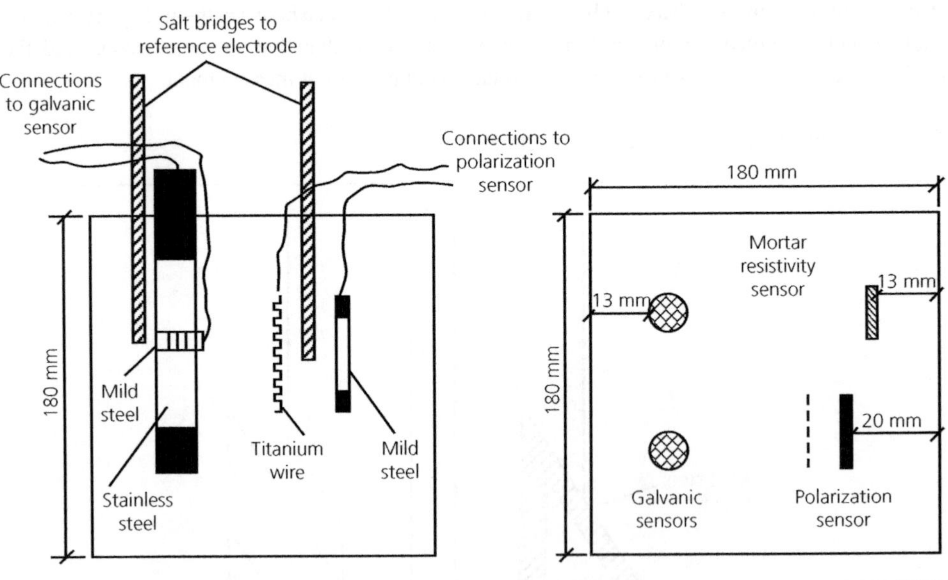

Table 1 Chemical analyses and ignition loss (wt %) of cements

	CaO	SiO$_2$	Al$_2$O$_3$	Fe$_2$O$_3$	SO$_3$	MgO	Na$_2$O	K$_2$O	Ignition loss
OPC	64·5	20·0	5·3	3·4	3·0	1·2	0·10	0·78	0·9
ggbs	41·0	34·0	12·0	1·0	1·0	8·5	0·50	0·30	–
PFA	1·9	51·1	28·2	8·1	0·8	1·3	0·96	3·71	2·8

Fig. 4 Controlled environment tank for accelerated carbonation studies

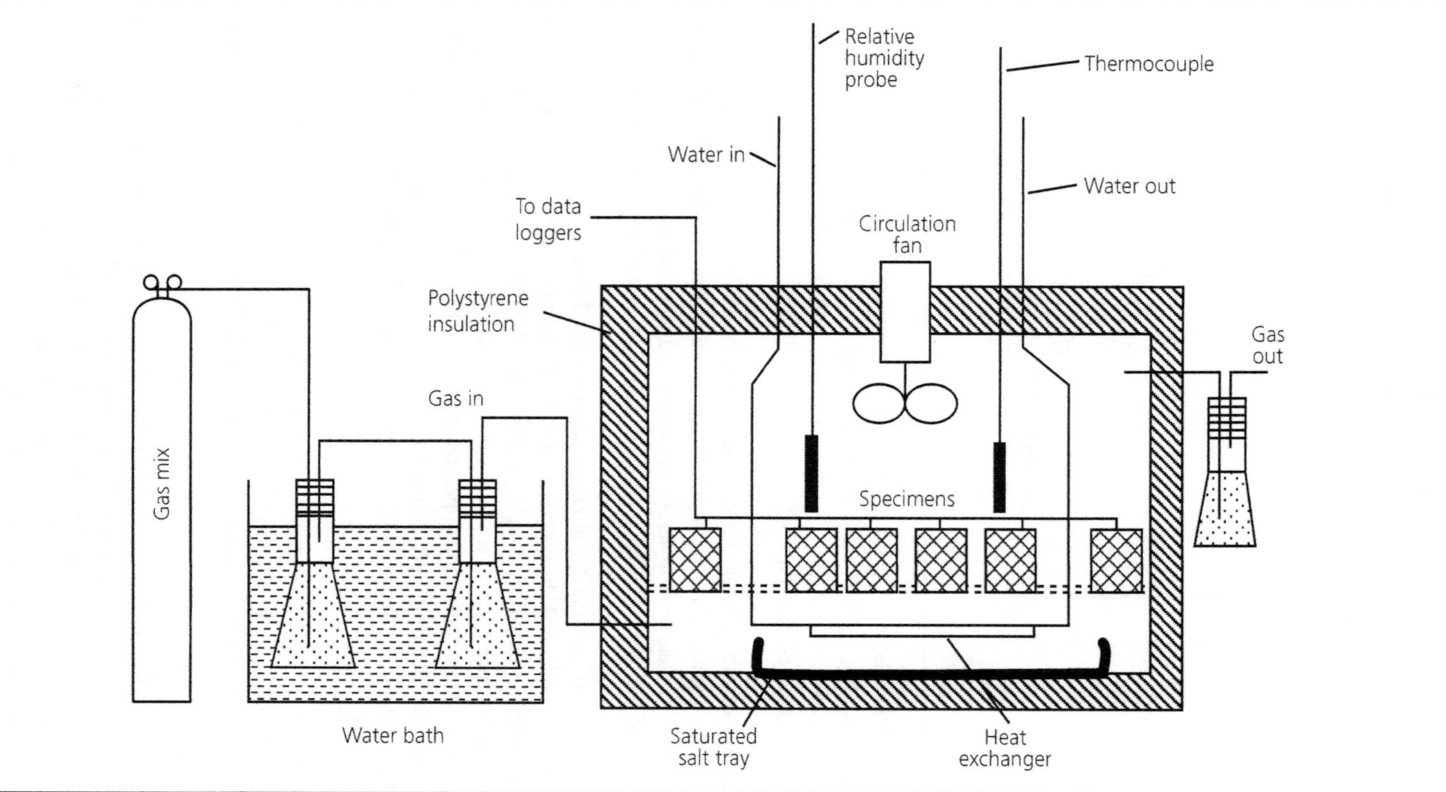

Fig. 5 Rate of carbonation for the mortar cubes

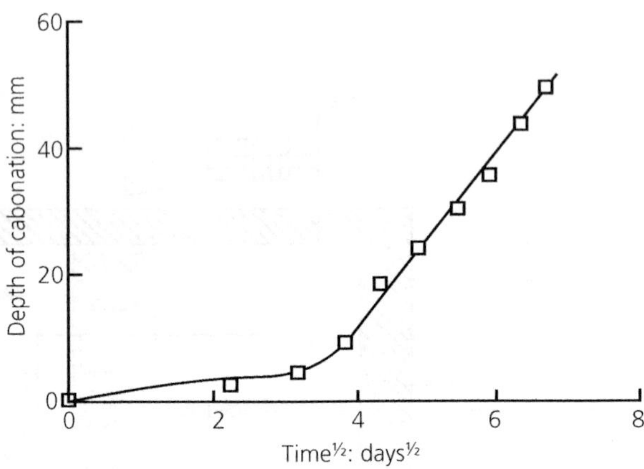

established that the kinetics of carbonation of the mortar cubes in an atmosphere of 5%CO_2 −20%O_2−75%N_2, at 25°C and 65% RH conformed to a parabolic relationship, and that carbonation at depths of cover to the sensors would be achieved within conveniently short experimental times (see Fig. 5).

In the case of the galvanic sensor, the potentials of the stainless steel and mild steel were measured by means of a Solartron 7060 digital voltmeter. Galvanic current was determined by measuring the potential drop across a resistor of suitable value (much lower than the electrolytic resistance of the mortars) after it had shorted the mild steel-stainless steel couple for 5 min.

For both types of sensor, R_p values for the uncoupled electrodes were measured with an Amel potentiostat (model 551) by potentiostatically shifting the mild steel potential by 10–15 mV and recording the current after 60 s. R_{unc} was measured using a signal generator to produce a 40 Hz, 30 mV peak-to-peak square wave, the positive feedback facility on the Amel potentiostat, and an oscilloscope to detect undamped oscillations. This value of R_{unc} was subtracted from the experimentally determined R_p to give a corrected value. From the corrected values of R_v thus obtained, corrosion intensities (I_{corr}) were calculated by means of the Stern-Geary equation[13]

$$I_{corr} = B/R_p$$

where $B = (\beta_a \times \beta_c)/2 \cdot 3(\beta_a + \beta_c)$. The anodic and cathodic Tafel coefficients β_a and β_c were both assumed to be 120 mV/decade, which gave $B = 26$ mV. This result has been found to agree acceptably with gravimetric determinations for steel suffering active corrosion in concrete.[14]

Results

An example of the electrochemical response of a galvanic sensor embedded in an OPC mortar is shown in Fig. 6. Accelerated carbonation of the mortars started when samples were 62 days old (shown as zero time in Fig. 6) and continued for a period of 25 days. During the carbonation period, the RH in the tank was difficult to control and fluctuated between 63% and 79%, probably as a result of release of moisture during carbonation. The RH was then altered step-wise for successive periods as indicated at the top of Fig. 6(a).

Fig. 6 Changes in electrochemical response of a galvanic sensor embedded in an OPC mortar: (a) potentials (mV) of the mild and stainless steels; (b) galvanic current between the mild and stainless steels; (c) corrosion intensity of the mild steel calculated from measurements of R_p and R_{unc}

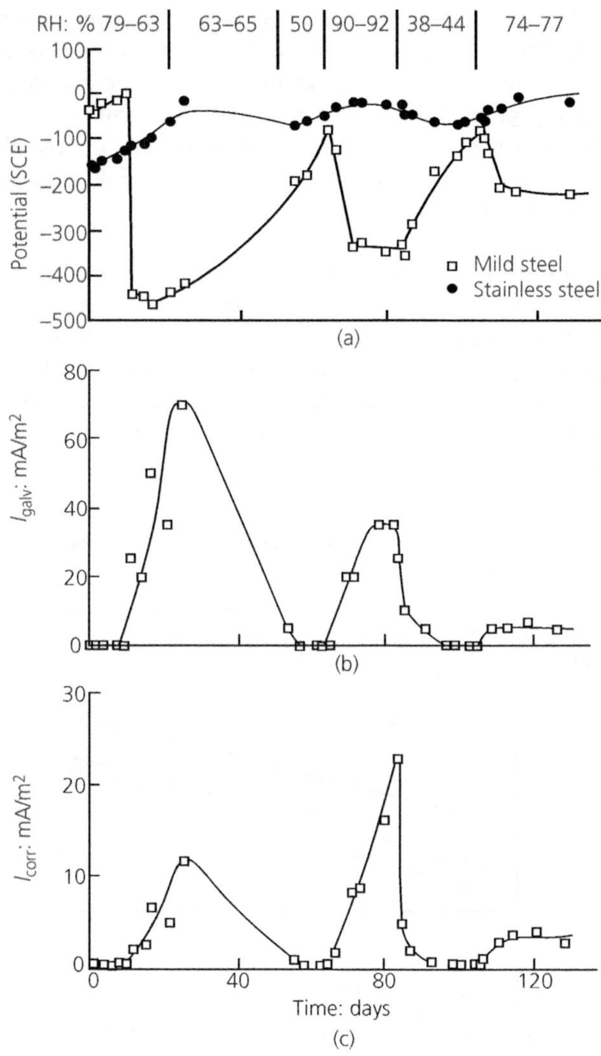

Carbonation penetrated to the level of the embedded sensors after an estimated 10 days, at which time there was a marked decrease in the potential of the mild steel accompanied by increases in I_{galv} and I_{corr}. Thereafter the electrochemical behaviour followed changes in RH, so that a lower RH (e.g. 50% or 38–44%) produced smaller galvanic currents and corrosion intensities with less negative mild steel potentials, while the converse was true for a higher RH (e.g. 90–92% or 74–77%). Although fluctuations in stainless steel potential were observed, these were generally small as compared to the changes in potential of the mild steel.

The reproducibility of results from galvanic sensors in identical mixes was generally very good; essentially similar trends to those shown in Fig. 6 were found for blended cement samples and for mixes containing 0.4 and 1.0% Cl. Using average values of I_{galv} from a number (~ 4) of sensors, regular trends in behaviour could be determined. For example, from the values of I_{galv} taken towards the end of an RH cycle when conditions had stabilized, it was possible to demon strate the sensitivity of the galvanic sensor to changes in RH (a regular increase in I_{galv} with increasing RH) and the presence of chlorides (an increase in I_{galv} at any given RH), as shown in Fig. 7(a), and the influence of blended cements (Fig. 7(b)). When the values of I_{galv} shown in Fig. 7 were plotted against the corresponding values of I_{corr}, a linear relationship (slope = 3.4, correlation coefficient = 0·97) was found, as shown in Fig. 8.

Changes in E_{corr} and I_{corr} for the polarization resistance probe embedded in the same mortar as the galvanic sensor are shown in Fig. 9. The initial marked decrease in E_{corr} occurred about 5 days later than in the case of the galvanic sensor, due to the greater depth of cover. Thereafter, changes in E_{corr} and I_{corr} followed exactly the same trends as for the galvanic sensor. It should be noted, however, that values of I_{corr} for this probe were lower than those for the galvanic sensor. This was also noticed when comparing the responses of the two devices in other mixes.

Fig. 7 Variation in corrosion intensity of steel with relative humidity in carbonated mortars: (a) influence of chlorides; (b) influence of blended cements

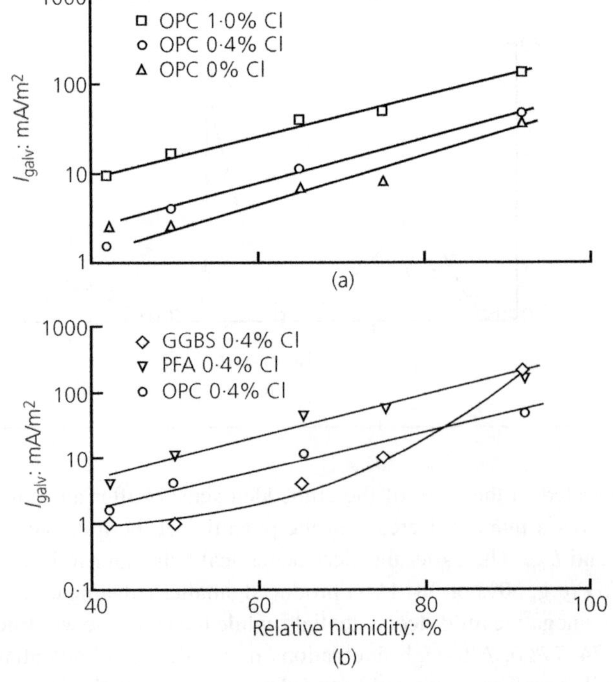

Fig. 8 Correlation between corrosion intensity of the mild steel and the galvanic current between the mild and stainless steels for samples of various mix designs

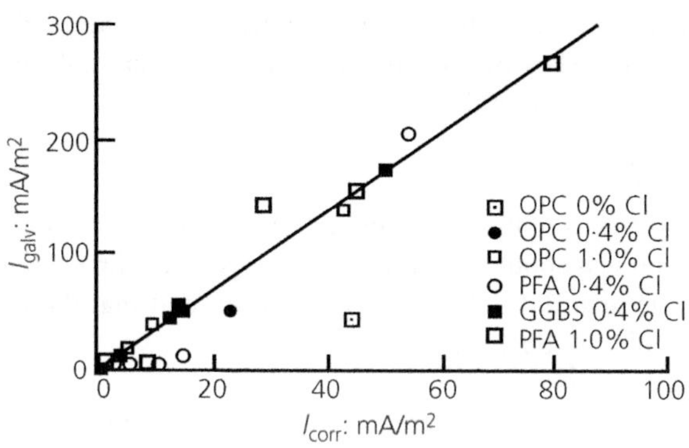

Fig. 9 Electrochemical response of a polarization resistance sensor embedded in an OPC mortar: (a) corrosion potential of the mild steel; (b) corrosion intensity of the mild steel

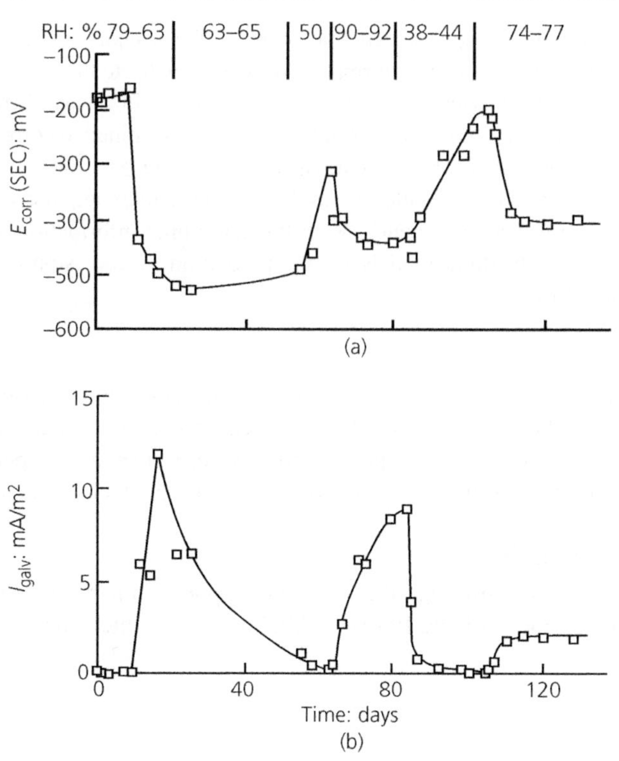

Discussion

Comparing the results shown in Figs 6(b) and 6(c), it is evident that the galvanic response I_{galv} of the simple sensor is related to the corrosion intensity I_{corr} of mild steel as measured by the more elaborate electrochemical technique of polarization resistance determination. If the corrosion rate were controlled by oxygen diffusion to cathodic sites, the expected relationship between I_{galv}, and I_{corr} would be of the form

$$i_{galv} = I_{corr} \times (A_c/A_a)$$

where A_a and A_c are the areas of the anode and cathode respectively.[9] In the present study the cathode/anode area ratio was about 200, whereas the slope of the line in Fig. 8 is only 3·4. This shows that the cathode area did not have a large effect, and that the above mechanism of corrosion rate control did not operate.

A more complex polarization resistance probe, of the sort shown in Fig. 2, confirmed the trends in I_{corr} found with the galvanic sensor. Results obtained with the former device have been used as a means of more detailed elucidation of the mechanism of steel corrosion in carbonated concrete: these aspects of the work are reported elsewhere.[4] Although the probe illustrated in Fig. 2 has some advantage in terms of accuracy, it is less robust and requires more complex instrumentation than the galvanic sensor. It is therefore considered that the latter device is likely to be more suitable for the prediction of relative corrosion rates in the majority of long-term exposure tests and similar field applications.

In field studies it should be appreciated that, even if the sensors are embedded in close proximity to the reinforcing bars, the information obtained relates to the sensor itself and not to the main steel. It cannot be assumed that the reinforcement will always behave in a similar manner to the sensor: a sensor 'constant' would have to be determined to correlate the values of I_{galv} and weight loss measurements of the adjacent reinforcing bar, and a number of sensors would have to be installed to monitor a single structural member, the optimum number required being determined by experience. With this additional information, galvanic sensors should be helpful in establishing models for the prediction of corrosion behaviour of reinforced concrete structures.

Conclusions

A galvanic sensor has been devised which responds to changes in the corrosion rate of steel in carbonated mortars. The sensor is easy to construct, and its associated measuring instrumentation is simple. It is robust and has potenial for development as an inexpensive method for long-term continuous non-destructive corrosion monitoring of steel in real structures.

Acknowledgements

The authors gratefully acknowledge the financial support of the Science and Engineering Research Council and the Building Research Establishment, which enabled this work to be carried out.

REFERENCES

1. Richardson M. G. *Carbonation of reinforced concrete: its causes and management.* CITIS Ltd, Dublin, 1988.
2. Parrott L. J. *A review of carbonation in reinforced concrete.* British Cement Association, Slough, 1987, technical report C/1-0987.
3. Gonzalez J. A. *et al.* Corrosion of reinforcing bars in carbonated concrete. *Br. Corros. J.*, 1980, **15**, 135–139.
4. Glass G. K. *et al.* Factors affecting the corrosion rate of steel in carbonated mortars. *Corros. Sci.*, 1991, **32**, No. 12, 1283–1294.
5. Dawson, J. L. *et al.* (Page C. L. *et al.* (eds)). Electrochemical methods for the inspection and monitoring of corrosion of reinforcing steel in concrete. *Corrosion of reinforcement in concrete.* Elsevier Applied Science, London, 1990, pp. 358–371.
6. Dhir R. K. *et al.* Measurement of reinforcement corrosion in concrete structures. *Concrete,* 1991, **25**, No. 1, 15–19.
7. Cavalier P. G. and Vassie P. R. Investigation and repair of reinforcement corrosion in a bridge deck. *Proc. Inst. Civ. Eng.*, Part 1, 1981, **70**, 461–480.
8. Page C. L. (Masters L. (ed.)) Barriers to the prediction of service life of metallic materials. *Problems in service life prediction of building and construction materials.* Martinus Nijhof, The Hague, 1985, E-95, pp. 59–74.
9. Mansfeld F. (Parkins R. N. (ed.)) New approaches to atmospheric corrosion research using electrochemical techniques. *Corrosion processes.* Applied Science Publishers, London, 1982, pp. 1–76.
10. Beeby A. W. Development of a corrosion cell for the study of the influence of environment and concrete properties on corrosion. *Concrete 85–the performance of concrete and masonry structures, Brisbane,* 1985.
11. Naish C. *The development of a galvanic probe for monitoring the corrosion of steel in concrete.* Transport and Road Research Laboratory, Crowthorne, 1989, TRRL contractors report 152.
12. Tuutti K. *Corrosion of steel in concrete.* Swedish Cement and Concrete Research Institute, Stockholm, 1982, report F04.
13. Stern M. and Geary A. L. Electrochemical polarisation I. A theoretical analysis of the shape of polarisation curves. *J. Electrochem. Soc,* 1957, **104**, No. 1, 56–63.
14. Andrade C. and Gonzalez J. A. Quantitative measurements of corrosion rate of reinforcing steels embedded in concrete using polarisation resistance measurements. *Werks. Korros.*, 1978, **29**, 515–519.

Dhir and Paine
ISBN 978-0-7277-6457-7
https://doi.org/10.1680/icetsc.64577.183
ICE Publishing: All rights reserved

Chapter 12

Corrosion monitoring of reinforced concrete beam using embedded cement-based piezoelectric sensor

Youyuan Lu
PhD Graduate, Department of Civil Engineering, Hong Kong University of Science and Technology, Clear Water Bay, Hong Kong

Jinrui Zhang
PhD Candidate, Department of Civil Engineering, Hong Kong University of Science and Technology, Clear Water Bay, Hong Kong

Zongjin Li
Professor, Department of Civil Engineering, Hong Kong University of Science and Technology, Clear Water Bay, Hong Kong

Biqin Dong
Professor, Guangdong Province Key Laboratory of Durability for Marine Civil Engineering, Key Laboratory on Durability of Civil Engineering in Shenzhen, School of Civil Engineering, Shenzhen University, Shenzhen, China

The corrosion of reinforcement is a well-recognised problem in the maintenance of concrete structures that can lead to reduction of the cross-section of the rebar, thus weakening the load-carrying capacity of the structure. Under ordinary design conditions, cracks are allowed to appear in concrete beams. However, the existence of cracks can speed up the corrosion rate of reinforcement significantly. In this study, the corrosion activities of reinforcement in concrete beams with or without cracks under accelerated corrosion conditions were monitored using embedded cement-based piezoelectric composite sensors and the acoustic emission (AE) technique. The AEs generated by the initiation of localised corrosion and subsequent concrete cracks were successfully detected and recorded by a home-programmed DEcLIN monitoring system. Based on frequency domain identification, useful information was extracted to explain the process of deterioration of concrete structures. It was found that cement-based piezoelectric sensors show a good capability in AE detection and the beam with cracks corroded much earlier than the case without cracking, consequently resulting in early deterioration.

Notation

$f(t)$	frequency domain discrete function of detected original AE signal
$f'(t)$	frequency domain discrete function of filtered AE signal
$G(t)$	elastodynamic Green's function for mortar
k_s	Kolosov constant of the rebar

$M(t)$	time domain function of the acoustic source
$T(t)$	response function of the monitoring system
T_a	time of onset of accelerated aerobic corrosion
T_c	time of onset of concrete damage phase
T_d	time when critical chloride concentration is reached
$V(t)$	detected AE signal waveform function in the time domain
α	reduction ratio of the diameter of the reinforcement (and the corrosion rate)
μ_c	shear modulus of the surrounding concrete
μ_s	shear modulus of the rebar
σ_θ	maximum shear stress induced by the volume expansion of corrosion products
ϕ	volume ratio of the final corrosion product over the original steel volume
ψ	identification threshold

Introduction

The corrosion of reinforcement corrosion is a major deterioration problem in concrete structures, especially in marine and coastal environments. Although both carbon-induced corrosion and chloride-induced corrosion are harmful to the reinforcement, chloride-induced corrosion is more dangerous due to its localised nature. Chloride ions can enter concrete structures by means of contaminants in the raw mixing materials during construction and can also be introduced by environmental exposure. It normally takes a relatively long time for ions to be transported to the location of the reinforcement and accumulate to a critical level through the concrete covers. However, the speed of the transportation process can be greatly increased due to the presence of microcracks and macrocracks when the concrete structure is under load. The coupled effect of loading and chloride-induced corrosion thus needs to be studied since it is in accord with the real situation of many concrete structures.

Acoustic emission (AE) monitoring has been used to detect the corrosion process of steel in concrete (Li *et al.*, 1998). It has been shown that AE technology has a unique advantage over other inspection methods since it could theoretically be generated by both the onset of the corrosion (Cho and Takemoto, 2004; Ohtsu and Tomoda, 2008) and the cracking of concrete. Therefore, with appropriate AE sensors, AE monitoring results will directly reflect the corrosion process of the reinforcement.

In this study, cement-based piezoelectric sensors were embedded into concrete structures for the purpose of AE detection; such sensors are quite suitable for the long-term monitoring of concrete structures because of their improved sensitivity and durability in concrete. A signal-based AE evaluation method can be used to qualitatively study the deterioration mechanics of reinforced concrete (RC) structures. However, given that corrosion is a complicated long-term physical and chemical process, the reliability of detected AE information has to be confirmed before performing signal-based AE evaluation. A frequency domain identification method is used to filter out fake AE signals based on the frequency domain analysis results of real signals (Leelalerkiet *et al.*, 2005; Li *et al.*, 1998) and the experimental results give a good explanation of the theoretical corrosion mechanisms.

Corrosion mechanisms of RC structures

Considering the situation of steel reinforcement in concrete, several phenomenological models have been proposed to represent the corrosion rate of reinforcement as a function of time.

Among them, Melchers and Li (2006) proposed a typical corrosion model for steel in water. Based on the fundamental reactions occurring on steel in an alkaline concrete environment, as shown in Equations 1 and 2, the corrosion process can be divided into four main phases: chloride and hydroxyl ion diffusion; accelerated aerobic corrosion; an aerobic corrosion stable phase; and an anaerobic corrosion phase.

The anodic reaction is

$$\text{Fe} \rightarrow \text{Fe}^{++} + 2e^{-} \tag{1}$$

$$\text{Fe}^{++} + \text{OH}^{-}, \text{O}_2, \text{H}_2\text{O} \rightarrow \text{Fe(OH)}_2,$$

$$\alpha\text{-FeOOH}, \gamma\text{-FeOOH}, \text{Fe}_3\text{O}_4, \alpha\text{-Fe}_2\text{O}_3$$

while the cathodic reaction is

$$0.5\text{O}_2 + \text{H}_2\text{O} + 2e^{-} \rightarrow 2\text{OH}^{-} \tag{2}$$

Since it may take quite a long time for anaerobic corrosion to take place – even longer than the time when observable severe concrete cracking has already occurred (Melchers and Li, 2006) – its destructive effect on steel in concrete was ignored. The onset of the accelerated aerobic corrosion phase in steel reinforcement is governed by the transportation properties of concrete and the thickness of the concrete cover (Ohtsu and Tomoda, 2008), which are described by Fick's law. However, when concrete structures are subject to service loadings, microcracks and macrocracks inevitably appear and provide additional convenient paths for the transportation of chloride ions towards the reinforcement (e.g. in the tensile area of a concrete beam cross-section). Therefore, the time needed for the surface chloride concentration to reach a critical level (the origin of x-axis in Figure 1) is greatly reduced, resulting in the time of onset of

Figure 1 Theoretical corrosion model of RC under the coupled effect of loading and chloride ingress. The y-axis represents the reduction of diameter of reinforcement due to corrosion, assuming that the reduction is uniform all round the circumference

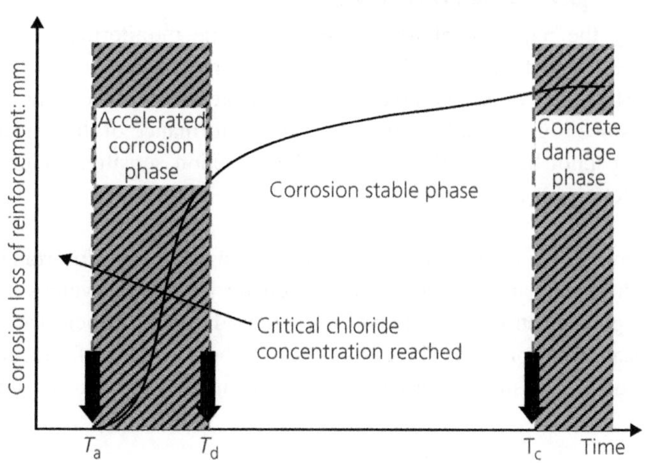

accelerated aerobic corrosion (T_a) occurring much earlier and more easily. Soon afterwards, the diffusion process will gradually slow down since corrosion products will inhibit the subsequent flow of chlorides and oxygen. The corrosion process then slows and stabilises from T_d, as illustrated in Figure 1. Knowing that the volume of the final corrosion products is more than six times the volume of steel, significant pressure will be induced around the circumference of the reinforcement. Assuming that the corrosion products grow uniformly around the circumference of the rebar, and by using the shrink-fit model (Li *et al.*, 1998), the maximum shear stress induced by the volume expansion of corrosion products is

$$\sigma_\theta = p = \alpha\left(\phi^{1/3} - 1\right) \frac{4\mu_c\mu_s}{2\mu_s + (k_s - 1)\mu_c} \tag{3}$$

where α is the reduction ratio of the diameter of the reinforcement (and the corrosion rate), ϕ is the volume ratio of the final corrosion product over the original steel volume, μ_c is the shear modulus of the surrounding concrete, μ_s is the shear modulus of the rebar and k_s is the Kolosov constant of the rebar.

Once σ_θ exceeds the critical level in the surrounding concrete at T_c, concrete cracks will occur and further develop. The concrete structures here were considered entering the concrete damage phase (see Figure 1). However, the onset of this phase is jointly determined by several parameters, such as the reinforcement corrosion loss rate and the shear capacity of the concrete material, which make it difficult to ascertain. Here, the corrosion process of steel in concrete under the coupled effect of loading and chloride ion attack is simply divided into four major phases with distinct characteristics

- critical concentration reached
- accelerated corrosion
- corrosion stable
- concrete damage.

Cement-based piezoelectric sensor

In order to satisfy the requirement for long-term reliable monitoring of the corrosion of concrete structures, embedded piezoelectric sensors were adopted in this work. Based on previous studies (Li *et al.*, 2002), compatibility of the piezoelectric sensing material and the matrix is critical since it greatly affects the detection performance of the embedded sensor. If the compatibility cannot be properly resolved, the detection sensitivity and signal-to-noise ratio are unsatisfactory for application.

Concrete is a dominant construction material in civil engineering. Unfortunately, there is a relatively large difference in the stiffness and acoustic impedance values of concrete and ordinary ceramic piezoelectric material and therefore traditional piezoelectric sensors are not suitable for application in concrete. Three issues need to be considered for tackling this issue – perfect coupling of the sensing contact interface, matching of the acoustic impedances and

Figure 2 Cement-based piezoelectric composite sensor

Figure 2 Cement-based piezoelectric composite sensor

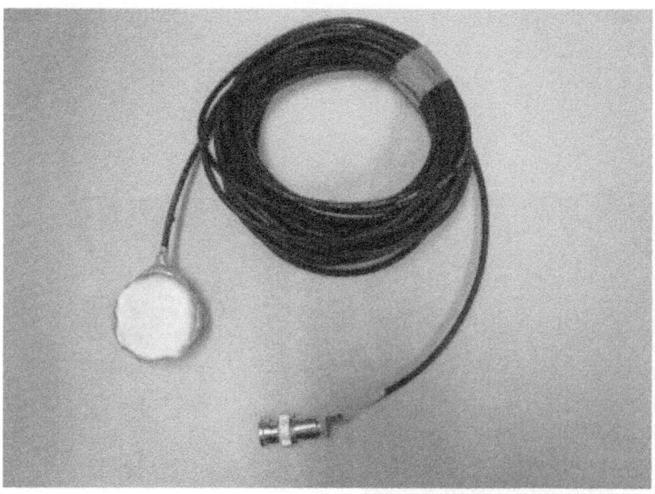

good mechanical compatibility with the matrix. It is possible to solve these problems using a cement-based piezoelectric composite material, which can be tuned to have an acoustic impedance value close to that of concrete matrix. Acoustic wave energy can then be maximally transferred to the sensing materials with little reflection at the medium interface, leading to high sensitivity of such material to AE. It was also found that the backing material supporting the piezoelectric sensor has a great influence on its damping characteristics. An active piezoelectric material of acoustic impedance similar to that of the backing material (concrete) will produce the most effective damping. Such a piezoelectric sensor will produce a flat and wide bandwidth when embedded into concrete. A cement-based piezoelectric composite was thus used to fabricate the sensors used in this work (Figure 2). The sensor calibration curve, obtained by a standard Hsu pencil lead break (Higo and Inaba, 1991), is illustrated in Figure 3. Broadband frequency domain characteristics with high sensitivity have been obtained (Lu and Li, 2008; Lu *et al.*, 2010; Qin *et al.*, 2010).

Experiment

Wet–dry cyclic accelerated corrosion tests were performed on four RC beams of dimensions $2100 \times 115 \times 100$ mm. Two beams were subjected to both loading and chloride-induced corrosion; these are denoted as beam A and beam B or simply as 'loaded'. The other two beams were subjected to only chloride-induced corrosion; these are denoted as beam C and beam D or simply as 'no load'. Three longitudinal rebars of diameter 16 mm were placed inside each concrete beam at a distance of 30 mm from the concrete beam's top and bottom surfaces respectively. The mixture proportion for the concrete was $1:0\cdot4:1\cdot6:2\cdot2$ (cement:water:sand: aggregate) with 1% superplasticiser (ADVA 105). Two cement-based piezoelectric sensors

Figure 3 Calibration curve of cement-based piezoelectric sensor: (a) time domain response; (b) frequency domain response; the grey shaded area indicates the flat frequency response region

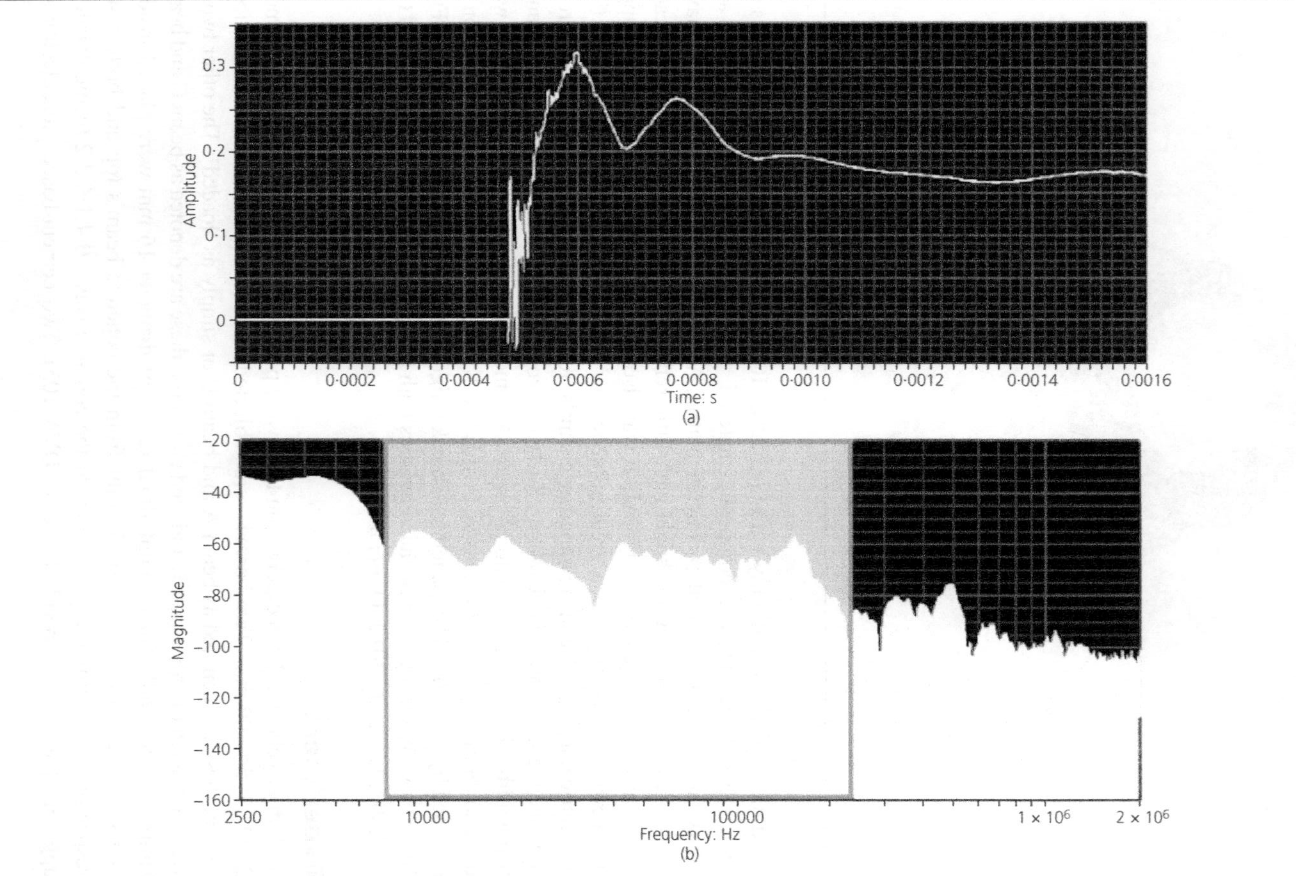

were positioned as shown in Figure 4. For beam A and beam C, the distance from the left-hand sensor to the left-hand beam end was 680 mm, while for beam B and beam D, this distance was 720 mm.

Beams A and B were subjected to four-point upwards bending successively during the tests (see Figure 5). Considerable transverse tensile cracks were expected to be induced between the loading points at the upper part of the beam (see Figure 6). According to BS 8110 (BSI, 1997), the applied load should not exceed 40% of the actual ultimate bearing capacity in order to avoid crack widths larger than 0.3 mm. The corrosion monitoring cycles normally started more than 2 weeks after the beams had been loaded. The existing crack patterns were found to be quite stable based on AE pre-monitoring. Almost no AE was detected before the wet–dry cycles. The creep behaviour of RC beams can be considered negligible under such circumstances (Bažant and Panula, 1978; Denarie *et al.*, 2006). Beams C and D were simply placed on the ground with no applied loading (see Figure 7).

Figure 4 Embedded cement-based piezoelectric sensor

Figure 5 Experimental set-up of the wet–dry accelerated corrosion of RC beam (dimensions in mm)

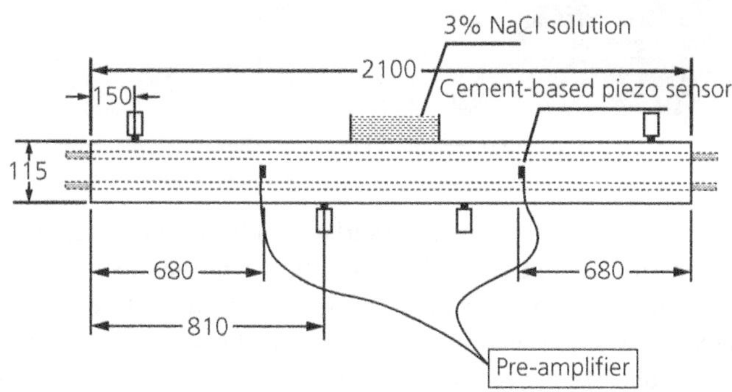

Figure 6 RC beams under the coupled effect of loading and corrosion

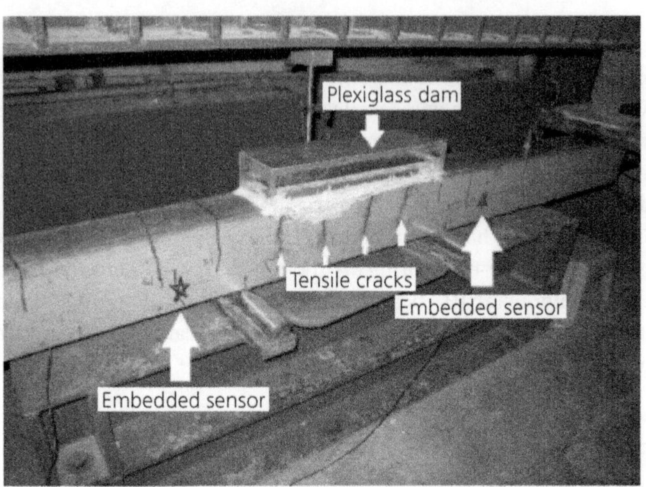

Figure 7 RC beams under the single effect of corrosion

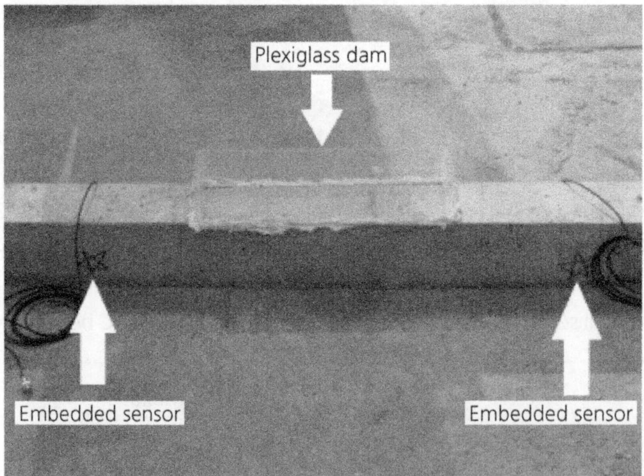

Plexiglass dams were placed on the centre top surfaces of the concrete beams (see Figures 5, 6, 7) to hold an appropriate amount of 3% NaCl solution. Chloride ions consequently diffused towards the reinforcement through the path of cracks. To reproduce the environmental effects of marine/coastal areas, 3-day wet and 4-day dry accelerated corrosion cycles using 3% NaCl solution were carried out. Beams A and C underwent a series of wet–dry monitoring cycles, while beams B and beam D had one. The results were used for comparison purposes. As already noted, both the corrosion activity of the reinforcement (Cho and Takemoto, 2004; Li *et al.*, 1998; Ohtsu and Tomoda, 2008) and concrete cracks can generate AE. The AE signals detected by the embedded sensors were modulated by preamplifiers (gain 40 dB, bandwidth 30 kHz–1 MHz) and transmitted to a home-programmed DEcLIN system. Useful information, such as accumulated AE event number (AEN) and the signal-based characteristic of detected AEs, was then collected for post-analysis and comprehensive study of the corrosion process of concrete under the coupled effect of loading and chloride ions.

Half-cell potential measurements according to ASTM C876-91 (ASTM, 1999) were also carried out simultaneously as a comparison and supplement. As described in the analysis guideline, it is empirically expected that if the half-cell potential reading obtained by using Cu/CuSO$_4$ standard reference electrodes is more negative than -0.35 V, there is a 90% probability of corrosion. However, it is frequently reported that the half-cell monitoring technique is difficult to ensure reliability when used alone during corrosion monitoring and the results obtained may be misleading (Leelalerkiet *et al.*, 2005).

Frequency domain identification

Acoustic emission monitoring is, up to now, one of the most reliable methods available for the detection of early stage corrosion. However, it would be expected that detected AE signals could result from not only the steel corrosion and concrete cracks but also from background noise from, for example, electromagnetic interference, unstable power supply and surrounding ultrasonic waves. Since reinforcement corrosion in concrete is a long-term process, there is a very high probability that background noise will contaminate real AE signals to a certain extent depending on the site conditions. Proper real-signal identification steps thus have to be taken before post-analysis.

Due to their distinct generation mechanism, it is feasible to make use of signal-based characteristic evaluation to distinguish real AE signals from fake AE signals. Li *et al.* (1998) found that signals emitted during corrosion have a frequency component distribution from 275 kHz to 350 kHz. It has also been reported that the AEs from concrete cracks have an average frequency range of 10–500 kHz (Li and Shah, 1994; Lu *et al.*, 2010; Reymond, 1980). These observed AE frequency ranges have a close relationship with the resonant frequency and the operating frequency range of the piezoelectric sensor employed. Given that cement-based piezoelectric sensors have a broad and flat frequency domain response (as discussed previously), detected signals with frequency components mainly distributed in the range of 30–600 kHz are regarded as real AE signals. An identification threshold (ψ) evaluates the proportion of signal frequency content distributed within the range of 30–600 kHz. Based on practical experience, any identified AE signal with a value ψ higher than 0·7 is classified as a real AE signal (Figure 8), while others are classified as fake (Figure 9). The identification threshold is given by

$$\psi = \left(\frac{\sum |f'(t)|^2}{\sum |f(t)|^2} \right)^{1/2} \tag{4}$$

where $f'(t)$ represents the frequency domain discrete function of filtered AE signals (30–600 kHz five-order band-pass filter) and $f(t)$ represents the frequency domain discrete function of detected original AE signals.

Figure 8 Identified real AE signal

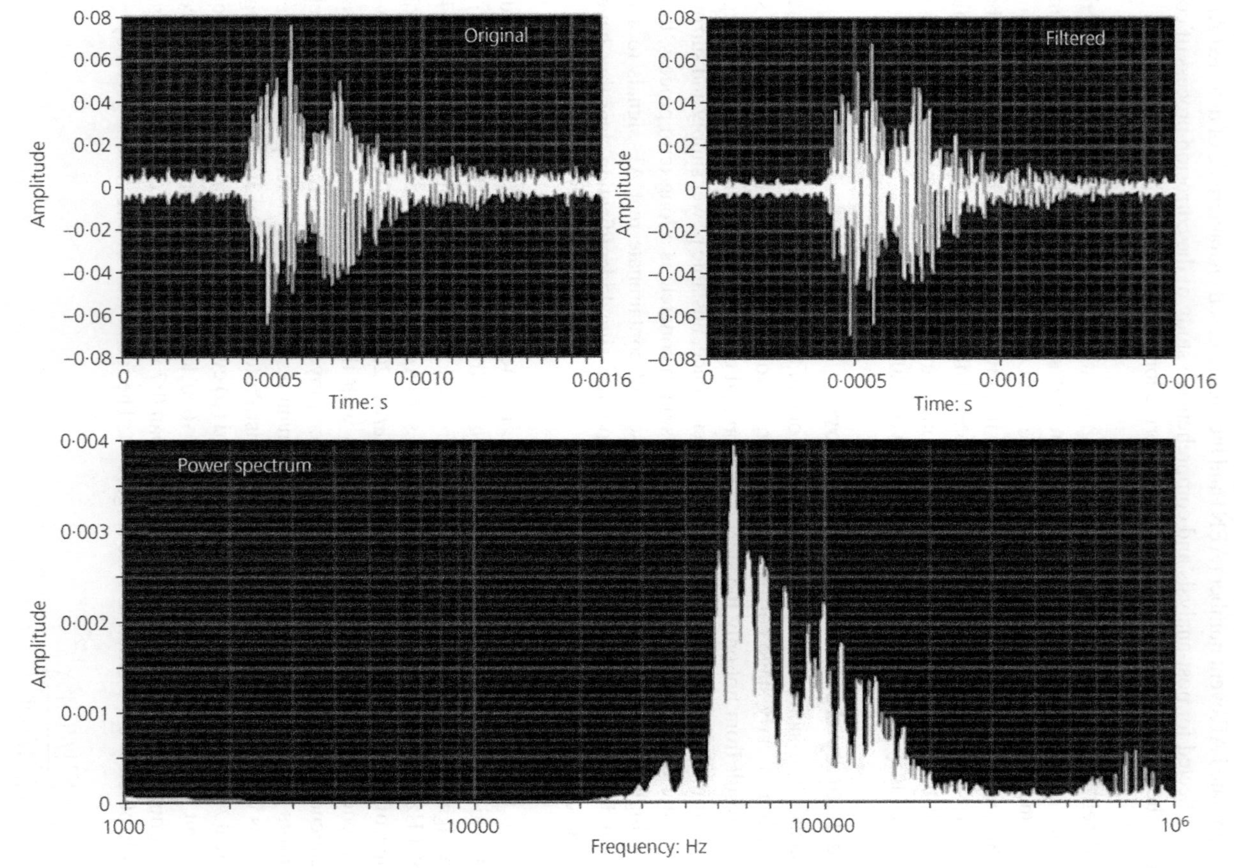

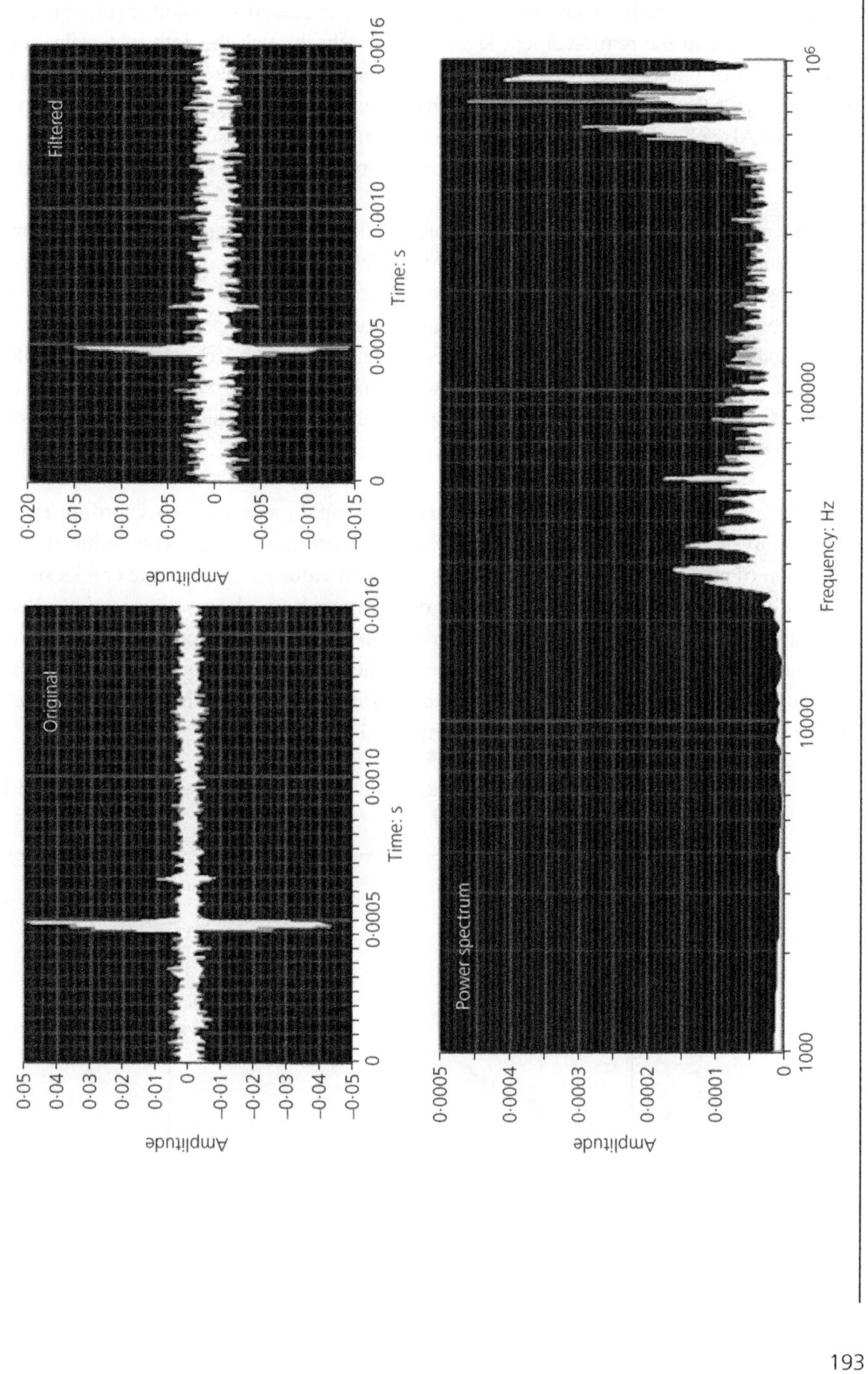

Figure 9 Identified fake AE signal

Results and discussion

With the AE signals being continuously detected during all wet–dry cycles, Figure 10 shows that two distinct AE active periods occurred in the second and sixth wet–dry cycles in the case of beam A. Given that tensile cracks of width no more than 0·3 mm were fully developed right below the plexiglass dam area due to applied static loading (see Figure 6), the chloride solution is supposed to have diffused to the reinforcement level immediately through the path of tensile cracks. The concentration of chloride ions began to increase continuously, accompanying $(OH)^-$ ions being leached out in the surrounding concrete, which resulted in AEs emerging promptly at the first cycle. Hence, the AE activity at this time mainly came from highly localised pitting corrosion activities, and the corrosion process then entered the critical concentration phase (Ohtsu and Tomoda, 2008).

At the second cycle, corrosion activities appeared in an accelerated manner and a significant amount of AE signals were obtained. The AEN (Figure 10) detected in the second cycle is some two to three times larger than that in the first cycle. Therefore, from the second cycle onwards, the corrosion process can be considered to be in the accelerated aerobic corrosion phase as illustrated in Figure 1. Theoretically, corrosion products quickly grew all around the rebar exposed to the tensile cracks. After that, AE activities seem to gradually slow down because of the build up of corrosion products, which retard oxygen and chloride ion transportation to the corroding surface of the steel. The corrosion rate thus decays to a stabilised level, as indicated by the corrosion stable phase in Figure 1.

After three further wet–dry cycles, AEN increases again at cycle 6. According to the theoretical analysis of the corrosion process, when the maximum shear stress induced by volume expansion of corrosion products overcomes a critical value of σ_c, concrete cracks surrounding the reinforcement develop. The increment of the maximum shear stress is controlled by the reduction ratio of the diameter of the reinforcement α due to corrosion, assuming the reduction in diameter is uniform along the circumference. Therefore, it is reasonable to relate AE activity to concrete cracking when the volume of corrosion products reaches a critical level after a certain time period; this is the concrete damage phase in Figure 1.

Figure 10 AE monitoring and beam deflection results during wet–dry cycles of loaded beam A (LVDT: linear variable differential transformer)

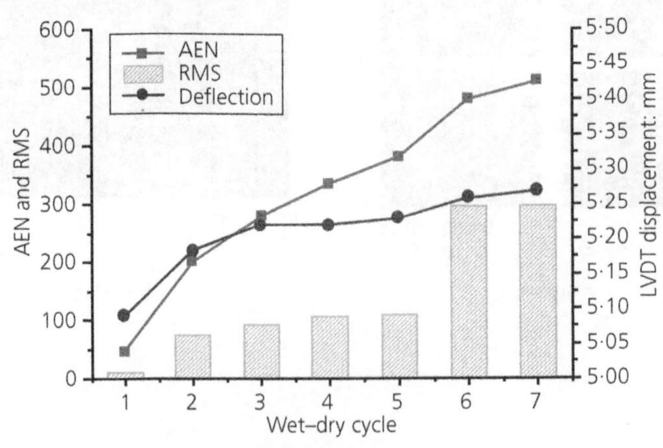

The observed AE phenomenon is very similar to that of beam B (Figure 11), which was under the same loading and corrosion conditions. Figure 11 clearly shows that the accelerated corrosion phase occurred in the second cycle. Cycles 3 to 8 belong to the corrosion stable phase, characterised by relative limited AEN being detected, during which corrosion products grew at a decreased rate. From cycle 9, the damage phase occurred, with an accelerated burst of AEN. It would seem that microcracks started to develop and propagate around the corroded reinforcement, as explained for the case of beam A. The detected signal RMS energy (RMS energy is a statistical measure of the energy magnitude of a varying quantity, such as detected AE signals) in the concrete damage phase (cycle 6 in beam A and cycle 10 in beam B) exceeds that of the signals in cycle 2 by 1·55 times for beam A and 1·28 times for beam B. Furthermore, both sets of results indicate that the occurrence of the accelerated corrosion phase is almost time-invariant under the same external conditions, but the occurrence of the concrete damage phase is time-variant even under the same external conditions. This is because the initiation of concrete damage is controlled by several factors such as the shear capacity of the cast concrete, corrosion rate and so on. This makes concrete crack initiation a complicated function of time, which is hard to predict precisely.

Figures 10 and 11 also explain the deflection variation at the mid-span of the beams throughout the wet–dry corrosion cycles. The deflection increased by 3·3% and 4·8% before cycle 5 in beam A and cycle 7 in beam B respectively and then the deflection variation clearly vanished in the following cycles in both cases. Based on analysis of the AE results, severe localised corrosion is expected straight after the critical chloride ion concentration is reached. Localised pitting corrosion apparently affected the bonding between the reinforcement and the surrounding concrete. The tensile stress carried by the concrete was thus reduced, which consequently increased the elongation of the reinforcement at the crack mouth. The localised pitting corrosion also weakened the stiffness of the reinforcement by reducing its cross-sectional area, causing the variation in deflection at mid-span of the beams while obvious deflection variation disappeared afterwards. It seems that concrete microcracks surrounding the corroded reinforcement did not have a significant effect on beam deflections and, conversely, proved that the pitting corrosion activity in this stage slowed down significantly

Figure 11 AE monitoring and beam deflection results during wet–dry cycles of loaded beam B

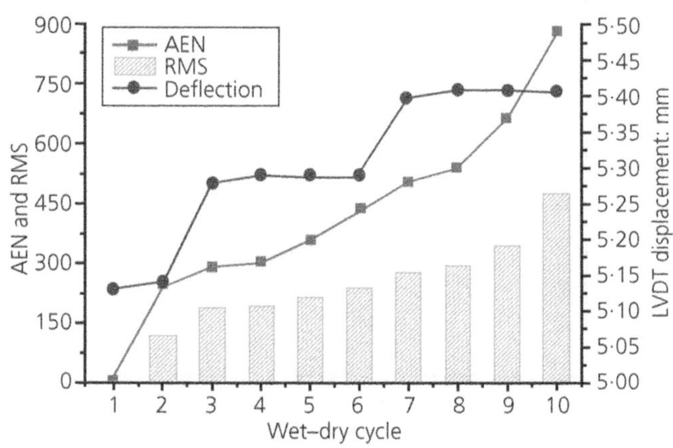

and stabilised. However, when major cracks caused by corrosion product growth eventually formed, spalling of concrete occurred, with fresh reinforcement exposed and consequently corroded. Pitting corrosion and accelerated corrosion will then appear again, but this is not the focus of the current study.

Figures 12 and 13 show the AEs detected under the coupled effect of loading and corrosion (Figure 6) and the effect of corrosion only (Figure 7). It is quite clear that, compared with the loaded beams, the AEs detected in the unloaded beams during the wet–dry cycles are limited and

Figure 12 AE monitoring during wet–dry cycles of beam A (loaded) and beam C (no load)

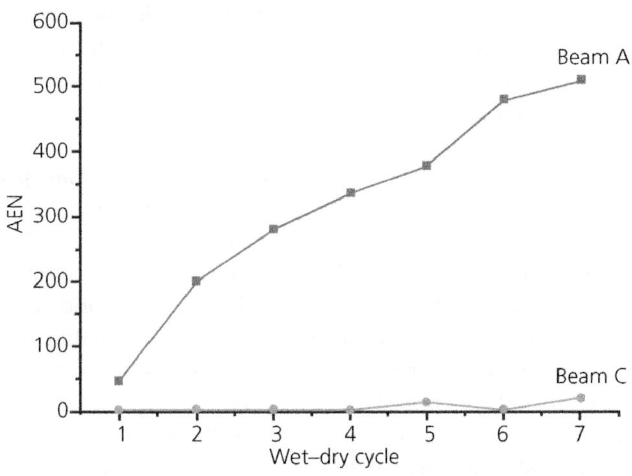

Figure 13 AE monitoring comparison during wet–dry cycles of beam B (loaded) and beam D (no load)

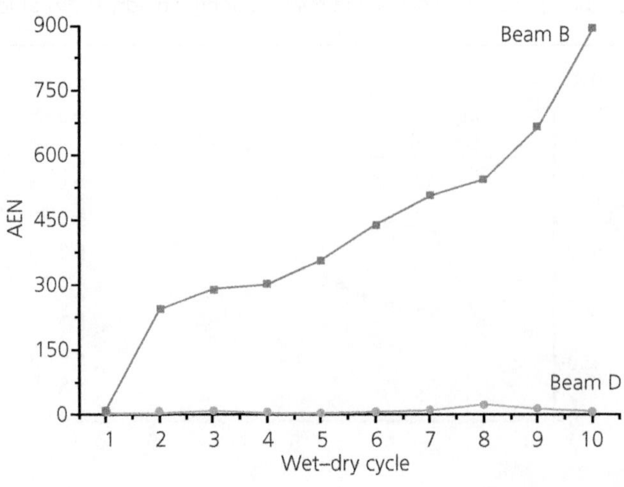

it can reasonably be concluded that no obvious corrosion activities occurred in beam C or beam D. As mentioned previously, the onset of corrosion in unloaded load RC beams is determined by Fick's law and the depth of concrete cover. Therefore, seven or ten wet–dry cycles are needed for chloride ion transportation. Clearly, the concrete cracks induced by loading greatly accelerate the onset of corrosion and the corresponding deterioration processes in RC beams.

Figures 14 and 15 show measured half-cell potential values before each wet cycle for both loaded and no load conditions. Figure 14 shows that no corrosion is implied in the first cycle

Figure **14** Half-cell potential of beam A (loaded) and C (no load)

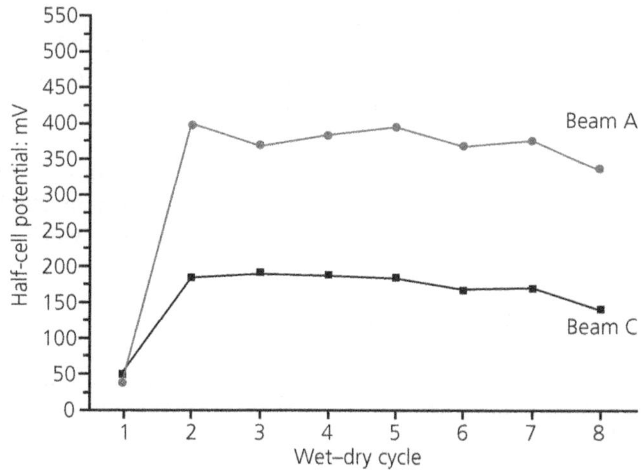

Figure **15** Half-cell potential of beam B (loaded) and D (no load)

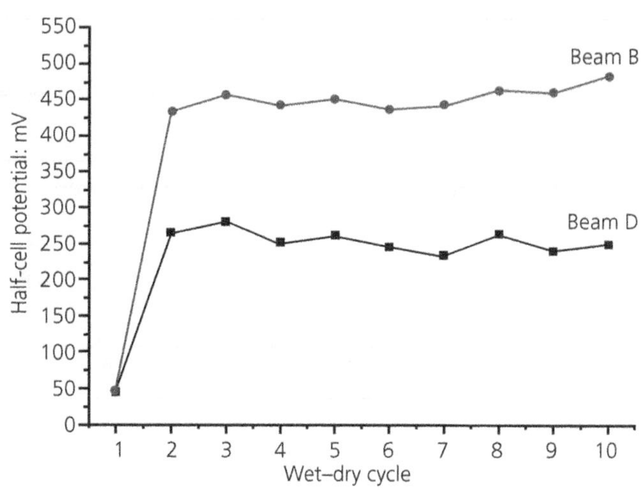

for either beam, while severe corrosion probably occurred from cycle 2 of beam A as the potential value quickly increased up over 350 mV according to ASTM C876-91. The potential measured then stayed at a relatively stable level at around 370 mV in beam A. However, the potential measured in corresponding beam C settled around 180 mV. This would indicate no corrosion in beam C.

Figure 15 shows that the potential variance of beam B is quite similar to that of beam A in Figure 14. At the first cycle, there was probably no corrosion in beam B. Then, the potential greatly increased, indicating that severe corrosion was probably initiated from cycle 2. The potential measured in beam B then stabilised at around 440 mV; the potential of corresponding beam D remained level at around 265 mV. Although the potential values of beam D are higher than those of beam C, both beams probably had no corrosion occurring during the wet–dry cycles. It is frequently reported that half-cell potential cannot be utilised as an accurate measurement of corrosion; here, it was used simply as a supplement to the AE results and it seems to match the analysis very well in this study.

Figures 16 and 17 show the estimated one-dimensional AE locations of beams A and B. The locations of the AE sources are mostly concentrated in the area directly below the dam area, with high chloride concentration and vertical tensile cracks, which agrees with the

Figure 16 Estimated AE locations during wet–dry cycles of beam A (triangles represent sensor positions)

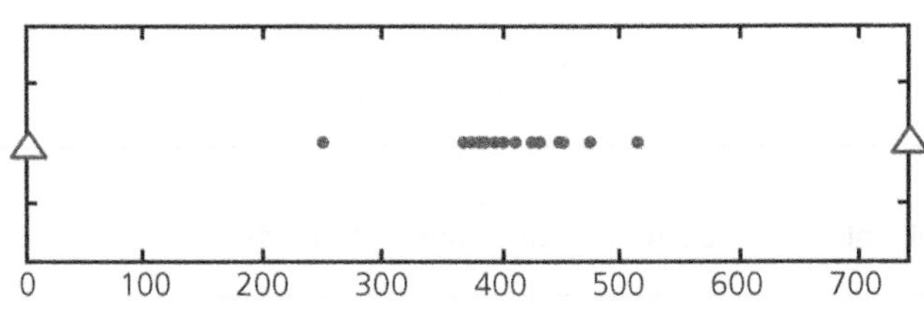

Figure 17 Estimated AE locations during wet–dry cycles of beam B (triangles represent sensor positions)

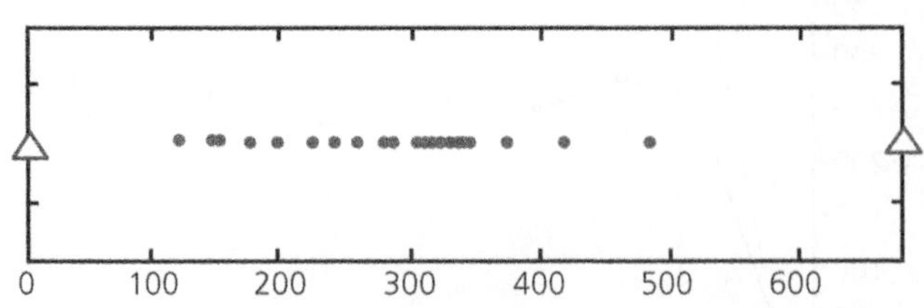

actual corrosion situation. Figures 18 and 19 show the corrosion of reinforcements after the wet–dry cycles. It can be seen that the beams subjected to the coupled effect of loading and chloride-induced corrosion (beams A and B) have severe localised corrosions in the area directly below the dam, while the beams with no loading display no notable corrosion of the reinforcements.

Figure 18 Comparison of reinforcement after wet–dry corrosion cycles: (a) mid-span of beam A; (b) mid-span of beam C

(a)

(b)

Figure 19 Comparison of reinforcement after wet–dry corrosion cycles: (a) mid-span of beam B; (b) mid-span of beam D

(a)   (b)

Figures 20 and 21 show the statistical distribution of the frequency spectrum centroid (SC) of AE signals acquired during the monitoring cycles evaluated by Equation 5

$$SC = \frac{\sum_{n=0}^{N-1} f(n)x(n)}{\sum_{n=0}^{N-1} x(n)} \tag{5}$$

in which $f(n)$ is the centre frequency of bin number n and $x(n)$ is the magnitude at the centre frequency of bin number n.

Figure 20 Cumulative proportion of frequency spectrum centroid in beam A

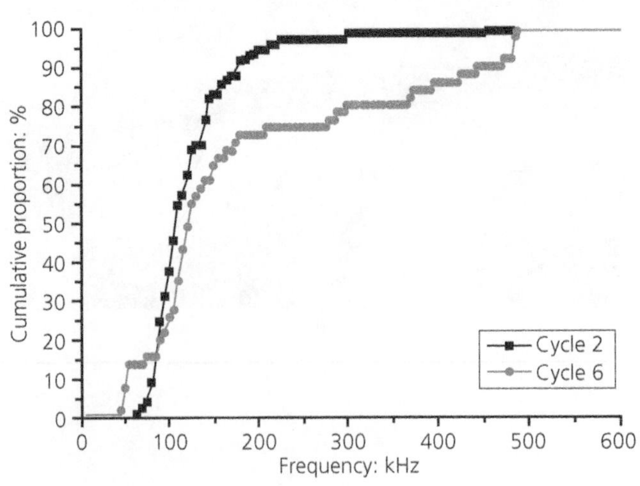

Figure 21 Cumulative proportion of frequency spectrum centroid in beam B

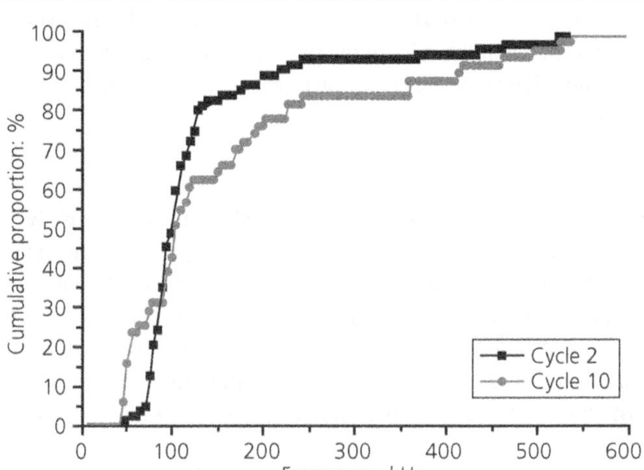

Given the direct physical link between acoustic source and the corresponding acquired AE signal as described by Equations 6 and 7, it is considered that the mechanics of AE sources during various corrosion stages would actively modulate the frequency content characteristics of the acquired AE signals accordingly, provided an unvarying propagation medium and monitoring system (Landis and Shah, 1995)

$$V(t) = T(t) * [G(t) * M(t)] \qquad (6)$$

where $V(t)$ is the detected AE signal waveform function in the time domain, $T(t)$ is the response function of the monitoring system, $G(t)$ is the elastodynamic Green's function for mortar, $M(t)$ is a time domain function of the acoustic source and * denotes a convolution integral.

$$V(S) = T(S) \cdot [G(S) \cdot M(S)] \qquad (7)$$

Equation 7 represents the Laplace transform of Equation 6 to illustrate the relationship between the detected AE signal and acoustic source in the S domain. Provided that $S = j\omega$, the frequency domain spectrum can be determined from Equation 7 by variable substitution.

Figures 20 and 21 show that, during cycle 2 for beam A (Figure 20) and beam B (Figure 21), the SC was principally concentrated in a narrow band around 100 kHz, whereas in 6 cycle for beam A and cycle 10 for beam B, the distribution of SC seems to be more widely spread, from 50 kHz to 550 kHz. The SC results indicate that, in the second cycle for both beams, the frequency contents of the AEs acquired are likely to be concentrated in a narrow band, while in cycle 6 for beam A and cycle 10 for beam B, the frequency contents of the AEs are widely scattered. These observations indicate that the AE signals acquired in these two stages were generated by AE sources from different mechanics. Compared with the analysis of AE RMS energy, two different AE sources were dominant in the AE peak cycles. It has been reported that the AE signals from corrosion lie in a narrow band (Li *et al.*, 1998), while the frequency ranges of AE signals from concrete cracks are widely scattered (Li and Shah, 1994; Maji and Sahu, 1994; Tanigawa *et al.*, 1980). By combining the trends of the evaluated AE parameters, it is reasonable to conclude that the AEs in the first AE peak stage are mostly due to corrosion initiation. In the second AE peak stage, concrete cracks begin to appear, which in turn affect the corrosion process significantly.

Conclusion

The corrosion processes of RC beams under the coupled effect of corrosive solution ingress and service loading were investigated using cement-based piezoelectric sensors and a home-programmed DEcLIN monitoring system. It was found that the obtained and evaluated AE parameters can be used to reflect and evaluate the corrosion process and the damage situation of concrete structures with satisfactory sensitivity and precision. Apart from laboratory testing, signal-based AE analysis has great potential for application to long-term corrosion monitoring of exposed RC structures. The following conclusions can be drawn from the current research.

- Embedded cement-based piezoelectric sensors are highly suitable for monitoring the corrosion process of RC structures. They are characterised by high sensitivity, good durability and a broadband and flat frequency domain response.
- Based on AE results, the onset of corrosion and concrete cracking can be identified. The results of half-cell potential and mid-span deflection variation measurements helped explain the corrosion mechanisms in the RC beams. It was found that the signal-based characters of AEs can be used to distinguish different AE source mechanics. This has great potential for use in interpreting different corrosion stages.
- Based on a comparison of beams with and without loading, it was found that the presence of cracks will clearly advance the onset of corrosion and the subsequent corrosion rate in RC beams. Therefore, RC beams under the coupled effect of loading and corrosive solution ingress will have an earlier deterioration problem.

Acknowledgements

Support from the China Ministry of Science and Technology (2009CB623200) and the Hong Kong Research Grant Council (N_HKUST637/9) is greatly appreciated.

REFERENCES

ASTM (1999) C876-91: Standard test method for half-cell potentials of uncoated reinforcing steel in concrete. ASTM International, West Conshohoken, PA, USA.

Bažant Z and Panula L (1978) Practical prediction of time-dependent deformations of concrete – part II: basic creep. *Matériaux et Constructions* **11(5)**: 317–328.

BSI (1997) BS 8110: Design and construction of reinforced and prestressed concrete structures. BSI, London, UK.

Cho H and Takemoto M (2004) Acoustic emission from rust in stress corrosion cracking. *Proceedings of European Working Group on Acoustic Emission*, BB 90-CD, pp. 503–510. See http://www.ewgae.eu/aboe.html (accessed 02/08/2013).

Denarie E, Cécot C and Huet C (2006) Characterization of creep and crack growth interactions in the fracture behavior of concrete. *Cement and Concrete Research* **36(3)**: 571–575.

Higo Y and Inaba H (1991) The general problem of AE sensors. *Acoustic Emission: Current Practice and Future Directions*. ASTM, West Conshohocken, PA, USA, ASTM STP 1077, pp. 439–457.

Landis E and Shah S (1995) Frequency-dependent stress wave attenuation in cement-based materials. *Journal of Engineering Mechanics ASCE* **121(6)**: 737–743.

Leelalerkiet V, Shimizu T, Tomoda Y and Ohtsu M (2005) Estimation of corrosion in reinforced concrete by electrochemical techniques and acoustic emission. *Journal of Advanced Concrete Technology* **3(1)**: 137–147.

Li Z and Shah S (1994) Microcracking localization in concrete under uniaxial tension: AE technique application. *ACI Materials Journal* **91(4)**: 372–389.

Li Z, Li F, Zdunek A, Landis E and Shah S (1998) Application of acoustic emission techniques in detection of reinforcing steel corrosion in concrete. *ACI Materials Journal* **95(1)**: 68–81.

Li Z, Zhang D and Wu K (2002) Cement-based 0–3 piezoelectric composites. *Journal of the American Ceramic Society* **85(2)**: 305–313.

Lu Y and Li Z (2008) Cement-based piezoelectric sensor for acoustic emission detection in concrete structures. *Proceedings of 11th International Conference on Engineering, Science, Construction, and Operations in Challenging Environments, Long Beach, CA, USA*. American Society of Civil Engineers, Reston, VA, USA, pp. 1–11.

Lu Y, Li Z and Qin L (2010) Signal-based AE characterization of concrete with cement-based piezoelectric sensors. *Computers and Concrete* **8(5)**: 563–581.

Maji A and Sahu R (1994) Acoustic emission from reinforced concrete. *Experimental Mechanics* **34(4)**: 379–388.

Melchers R and Li C (2006) Phenomenological modeling of reinforcement corrosive in marine environments. *ACI Materials Journal* **103(1)**: 25–32.

Ohtsu M and Tomoda Y (2008) Phenomenological model of corrosion process in reinforced concrete identified by acoustic emission. *ACI Materials Journal* **105(2)**: 194–199.

Qin L, Lu Y and Li Z (2010) Embedded cement-based piezoelectric sensors for AE detection in concrete. *Journal of Materials in Civil Engineering ASCE* **22(12)**: 1323–1327.

Reymond M (1980) Acoustic emission in rocks and concrete under laboratory conditions. *Proceedings of 2nd Conference on AE/MA in Geological Structures and Materials, University Park, PA, USA*. Trans Tech Publications, Clausthal, Germany, pp. 27–34.

Tanigawa Y *et al.* (1980) Frequency characteristics of acoustic emission waves of concrete. *Transactions of the Japan Concrete Institute*, pp. 129–132.

Dhir and Paine
ISBN 978-0-7277-6457-7
https://doi.org/10.1680/icetsc.64577.205
ICE Publishing: All rights reserved

Chapter 13
Performance of cement-based sensors with CNT for strain sensing

Carmen Camacho-Ballesta
Civil Engineer, PhD Student, Department of Civil Engineering, Universidad de Alicante, San Vicente del Raspeig, Spain

Emilio Zornoza
Associate Professor, Department of Civil Engineering, Universidad de Alicante, San Vicente del Raspeig, Spain

Pedro Garcés
Professor, Department of Civil Engineering, Universidad de Alicante, San Vicente del Raspeig, Spain

Strain-sensing functions of Portland cement pastes with different dosages of carbon nanotubes (CNT) are systematically studied. This is one of the relevant functions that multifunctional composites are able to develop when a conductive addition is included in the cement matrix. Strain sensing refers to the ability of a structural material to sense its own condition, such as strain or stress, through the piezoresistive behaviour (change of volume electrical resistivity when it is subjected to a force). CNTs, which possess exceptional electromechanical properties, are considered as a promising addition for practical applications such as structural health and load monitoring. Nevertheless, there is still an important challenge regarding CNT cement composite fabrication: achieving their effective dispersion. This point is addressed by means of a comparison between different physical and chemical dispersion methods. The technique used for that purpose is light-scattering particle size analysis. After selecting the optimal dispersing procedure, the effect on the strain-sensing properties of CNT reinforced cement pastes was studied for the following variables: CNT dosage, curing age, current intensity, loading rate and maximum stress applied. All these parameters are discussed taking gauge factor value as the reference.

Notation

D	material's density
E	elastic modulus
l_0	initial length
M_d	dry mass after drying in oven at $105 \pm 5°C$
M_s	saturated mass
M_s^{hb}	saturated mass measured in hydrostatic balance
P	material's total porosity
R_0	initial electrical resistance

Δl	specimen's deformation
ΔR	change on electrical resistance
ε	unit strain
σ	stress

Introduction

In the last few years, advances in nanotechnology have been conducted to obtain new nanoscale fibres with outstanding electromechanical properties, specifically carbon nanotubes (CNT). Thanks to their very low electrical resistivity (5×10^{-8}–$2 \times 10^{-6}\,\Omega$m, which is similar or superior to that of copper (Kim *et al.*, 2014)) they can be considered a conductive admixture for cement, making the development of new multifunctional, high-performance, advanced sensing, cement-based nanocomposites possible.

The strain-sensing function refers to the ability of a structural material to sense its own condition, such as strain or stress (which simply relates to the strain in the elastic regime) (Chung, 2012). When subjected to stress/strain, their electrical properties change, expressing a linear and reversible response, called piezoresistive response. Previous work has indicated that the contact resistivity between the matrix and the conductive admixture is responsible for the piezoresistive phenomena (Wen and Chung, 2006). In a practical way, if a longitudinal compressive stress is applied, the electrical resistance in that direction will be reduced. However, if the material is in tension, a contrary effect will be produced, that is, an increase in the electrical resistance will be registered. As both effects are reversible in the material's elastic range, the electrical resistance comes back to its initial value once the load is removed. This sensing ability can be useful for structural vibration control, load monitoring and structural health monitoring (Galao *et al.*, 2014).

Few studies have been performed on the strain-sensing ability of cement-based composites with the addition of CNTs. Li *et al.* (2007) conducted experiments using acid-treated CNTs and untreated CNTs as reinforcement and concluded that both types can decrease the electrical resistivity and improve pressure-sensitive properties of cement composites. Azhari and Banthia (2012) studied two types of cement-based sensors, one with just carbon fibres and the other carrying a hybrid of both carbon fibres and nanotubes. They found that both types showed piezoresistive response, providing the hybrid sensors with a better quality signal and increased sensitivity over sensors carrying carbon fibre alone. Yu and Kwon (2009) also found a piezoresistive behaviour for cement pastes reinforced with multiwalled carbon nanotubes (MWCNTs). Han *et al.* (2009) investigated a self-sensing CNT/cement composite for traffic monitoring in laboratory tests and road tests. Results showed that under repeated compressive loading the electrical resistivity decreases upon loading and increases when unloaded, leading to the conclusion that this nanocomposite can detect vehicular loads through remarkable changes in electrical resistance. Later, Han *et al.* investigated the effect of the MWCNT content and water/cement ratio on the piezoresistive sensitivity of composites. They examined cement nanocomposites with amounts of MWCNT of 0·05, 0·1 and 1 wt% (Han *et al.*, 2012). Experimental results concluded that the composite with 0·1 wt% of MWCNT presents the best sensing property. Luo *et al.* (2011) used cured MWCNT reinforced cement-based composites with 0·1 wt% and 0·5 wt% MWCNT and registered good piezoresistivity results and strain sensitivity for both samples, although the trendline of fractional change in resistivity presented better stability for amounts of 0·5 wt%. More recently, Konsta-Gdoutos and Aza (2014) investigated the resistivity of cementitious composites reinforced with CNT and carbon nanofibres (CNF). They found the addition of CNTs and CNFs at an amount of 0·1 wt% and 0·3 wt% with respect to cement mass induced a decrease in electrical resistance, with the nanocomposites containing 0·1 wt% CNTs yielding better electrical properties. Furthermore, their strain-sensing results confirmed that nanocomposites, reinforced with 0·1 wt% CNTs and CNFs, exhibited an increased change in resistivity, which is indicative of the amplified sensitivity of the material in strain sensing.

Usually the strain-sensing sensitivity is measured using the gauge factor (GF), which can be defined as the fractional change in electrical resistance per strain unit. This parameter can be calculated as follows

$$\text{GF} = \frac{\Delta R/R_0}{\Delta l/l_0} = \frac{\Delta R/R_0}{\varepsilon} \tag{1}$$

where ΔR is the change in electrical resistance, R_0 is the initial electrical resistance, Δl is the specimen's deformation, l_0 is the initial length and ε is the unit strain.

The main problem of CNTs is their difficult dispersion because of the highly attractive van der Waals forces between the CNT particles that tend to cause agglomeration (Figure 1). The insufficient dispersion of CNTs has been cited as a key diminishing factor for the performance of CNT–cement composites, as poor dispersion of CNTs leads to the formation of many defects in the nanocomposite and limits the mechanical properties of the composites. On the contrary, better dispersion of the CNTs may result in a higher interfacial contact area between the CNTs and the matrix, as well as more evenly distributed stresses in the composites (Chen *et al.*, 2011). According to the literature, there are many different dispersion technologies, but all of them can be classified into chemical or physical techniques (Parveen *et al.*, 2013). The basic physical technique used for CNT is ultrasonication, which is often used in combination with other chemical processes. Chemical approaches are designed as either a covalent treatment (CNT functionalisation), as researched previously for other carbon materials (Catala *et al.*, 2011), or a non-covalent treatment (dispersing chemical groups are physically attached onto the CNT surface). However, it has been reported that functionalisation might introduce structural defects resulting in inferior properties of the treated CNTs (Coleman *et al.*, 2006). Luo *et al.* (2009) assessed the dispersion capability in aqueous solution of five surfactants: sodium dodecyl benzene sulfonate, Triton X-100, sodium deoxycholate, Arabic gum and cetyltrimethyl ammonium bromide. The best results were obtained from a combination of the first two with a mixing ratio of 3:1, respectively. Cement composites reinforced with MWCNT subjected to that treatment had balanced mechanical and electrical properties. Collins *et al.* (2012) also investigated the dispersion of CNTs with different dispersants: air entrainer, styrene butadiene rubber, polycarboxylates, calcium naphthalene sulfonate and lignosulfonate formulations, achieving the best results for polycarboxylate additives.

Therefore, it is essential to integrate dispersion techniques into the fabrication of the CNT–cement composites for more effective use of CNTs as reinforcements. This point has been addressed first in the present research. Then a systematic study of strain-sensing properties in CNT reinforced cement pastes was performed. The effect of several variables on this function

Figure 1 Scanning electron microscope EM images of (a) non-dispersed, as received, CNT and (b) dispersed CNT

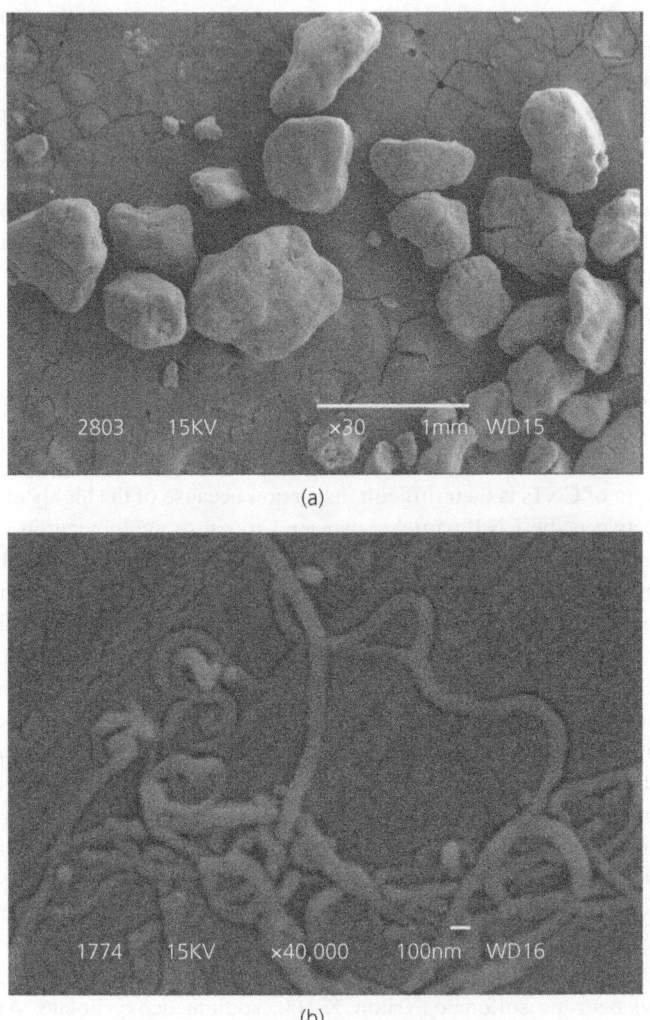

was studied, for example, CNT dosage, cement pastes curing age, current intensity, loading rate or maximum stress applied.

Experimental programme and materials

Materials and sample fabrication

Cement pastes were used for mechanical and physical tests (compressive and bending strengths, porosity and density) and strain-sensing tests, while fluid suspensions in water were prepared for dispersion trials. The materials used in this research were: Portland cement type

EN 197-1 CEM I 52·5 R; multiwall carbon nanotubes (MWCNT, Baytubes® C 70P), supplied by Bayer MaterialScience, S.A., whose characteristic properties are outlined in Table 1; distilled water; and Sikament-200 R, a commercial superplasticiser, supplied by Sika España. The following types of surfactants/dispersants were chosen because of their demonstrated solubilisation capability on MWCNTs (Collins *et al.*, 2012; Luo *et al.*, 2009)

- sodium dodecyl benzene sulfonate (SDBS)
- Triton X-100 (TX-100)
- polycarboxylate 1 (PC1), specifically Sika Viscocrete-20 HE, supplied by Sika España
- polycarboxylate 2 (PC2), specifically Glenium ACE 425, supplied by Basf Construction Chemicals España.

The water/cement ratio (w/c) for all pastes was 0·5 and CNT dosages were 0, 0·05, 0·10, 0·25 and 0·50% by cement mass. The dosages of superplasticiser were used for pastes according to previous research (Camacho-Ballesta *et al.*, 2014), for obtaining the same workability for all pastes. Thus the specific quantity of plasticiser was 0, 0·4, 0·5, 0·9 and 2·2% by cement mass, for CNT dosages of 0, 0·05, 0·10, 0·25 and 0·50%, respectively.

Carbon nanotube dispersions for pastes were done according to a previously checked method in polymer composites with CNF addition (Bortz *et al.*, 2011) and followed these steps: CNT and distilled water were poured into a high-shear mixer for 10 min and afterwards an ultrasound treatment was applied for 5 min using an ultrasound device model Hielschier UP200S. The resulting dispersion was mixed with cement and plasticiser in a laboratory mixer for 5 min. Pastes were fabricated in laboratory conditions: 20°C temperature and 65% relative humidity (RH). This dispersion method has been used successfully in CNF cement composites (Baeza *et al.*, 2013b; Galao *et al.*, 2012).

Table 1 Properties of Baytubes® C 70P multiwall CNT

Properties	Value	Unit
C-Purity	≥95	wt%
Free amorphous carbon	–	
Outer mean diameter	~13	nm
Inner mean diameter	~4	nm
Length	>1	μm
Bulk density	45–95	kg/m^3
Elastic modulus	3596	MPa
Tension at break	72·9	MPa
Elongation at break	10·7	%
Izod-impact at 23°C	103	J/m

For each dosage three prismatic specimens of $40 \times 40 \times 160$ mm^3 were fabricated according to AENOR Standard UNE EN 196-3:2005+A1:2009. The first specimen was used for mechanical and other characterisation tests (cured in water for 28 d), the two other specimens for piezoresistive sensing tests (cured in water until the first test date, that is 14 d, and afterwards in 100% RH condition).

Dispersion trials

A battery of suspensions was prepared by combination of different known methods for dispersing CNTs. They can be summarised as follows

- no treatment
- 10 min mixer
- 10 min mixer + 5 min ultrasound
- 0·25% of polycarboxylate 1
- 0·25% of polycarboxylate 2
- 0·25% of sodium dodecyl benzene sulfonate
- 0·25% of Triton X-100
- 5 min ultrasound + 0, 0·25, 0·50 and 1% of polycarboxylate 1 + ultrasound bath
- 5 min ultrasound + 0, 0·25, 0·50 and 1% of polycarboxylate 2 + ultrasound bath
- 5 min ultrasound + 0, 0·25, 0·50 and 1% of sodium dodecyl benzene sulfonate + ultrasound bath
- 5 min ultrasound + 0, 0·25, 0·50 and 1% of Triton X-100 + ultrasound bath.

For this study, as in previous dispersion studies, the dosage of CNT was established on 0·05% of cement mass (Collins *et al.*, 2012). The suspensions were prepared by manual agitation of CNTs in 500 ml of water with a glass bar followed by indicated treatments for each case. An additional ultrasound bath was also applied to the last four dispersions to guarantee the complete dispersant dissolution. Then a sample of about 10 ml was taken to be tested, keeping the suspension free of movements on a stable surface to observe the decantation process.

The instrument selected to evaluate the dispersions was a Coulter LS230, a light-scattering particle size analyser that measures particles from 0·04 μm to 2000 μm. It uses the diffraction of laser light by CNTs as the main source of information about particle size.

Mechanical properties and other characterisation tests

At 28-d age mechanical tests were conducted in laboratory conditions following the analogous procedure to AENOR Standard UNE EN 196-1:2005 for all pastes, with a ME-402/20 press machine (Servosis, S.A., Spain) to determine bending and compression strength for each dosage. Previously the non-destructive ultrasonic pulse velocity (UPV) data had been registered.

Porosity (P) and density (D) were calculated from at least two samples collected from the preceding mechanical tests. Apparent density was calculated as follows

$$D = M_{\mathrm{d}}/(M_{\mathrm{s}} - M_{\mathrm{s}}^{\mathrm{hb}})$$

where M_d is dry mass after drying in an oven at $105 \pm 5°C$, M_s is saturated mass and M^{hb}_s is saturated mass measured in hydrostatic balance.

Porosity was obtained as follows

$$P(\%) = (M_s - M_d)/(M_s - M^{hb}_s) \times 100$$

Porosity was also characterised by mercury intrusion porosimetry, using an Autopore IV 9500 V1·05 (Micromeritics Instrument Corporation).

Experimental set-up for measuring strain sensing

After the curing period samples were externally dried and silver electrically conductive paint (Pelco Conductive Silver 187) was applied around the perimeter at four interior planes which were parallel to the end surfaces. Additionally, four copper wires were wrapped around each silver painted perimeter, in order to form four electrical contacts, as needed for the four-probe method (Figure 2).

Each test consisted of at least four consecutive loading–unloading cycles (compressive). According to Equation 1, the gauge factor (GF) was calculated for each semi-cycle (i.e. only the loading or unloading part of each complete cycle) and for each completed cycle (considering both loading and unloading parts). The following parameters were systematically studied.

- CNT dosage: pastes with 0%, 0·05%, 0·10%, 0·25% and 0·50% CNT by cement mass.
- Curing age: samples were tested at 14 and 28 d. Between each test the samples were kept in a desiccator at room temperature and water saturated ambient (100% RH).
- Current intensity: 0·1, 1 and 10 mA.
- Maximum load of each cycle: 3, 6 and 9 kN loads, corresponding to stress values of 1·9, 3·8 and 5·6 MPa, respectively.
- Loading rates (velocity): 0·05, 0·10 and 0·20 kN/s for a 3 kN maximum load, and 0·10, 0·20 and 0·40 kN/s for 6 and 9 kN maximum loads.

Figure 2 Electrical contacts configuration: 1 and 2 are contacts for current input and 3 and 4 are contacts for voltage measurement

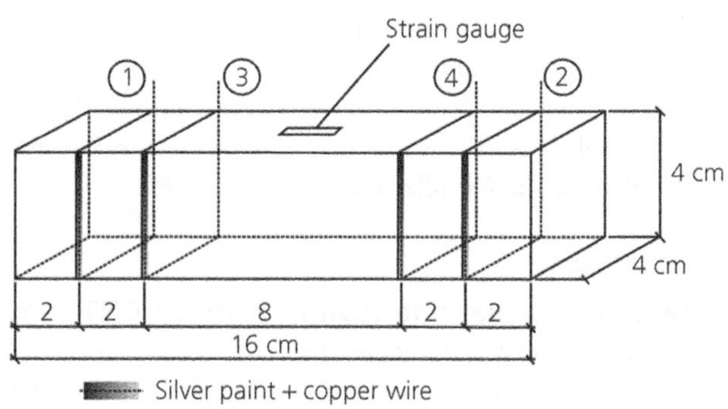

Figure 3 Diagram of strain-sensing test set-up. Current source and multimeter make it possible to obtain electric resistance values, while extensometer and press provide mechanical data for unit strain and applied stress, respectively

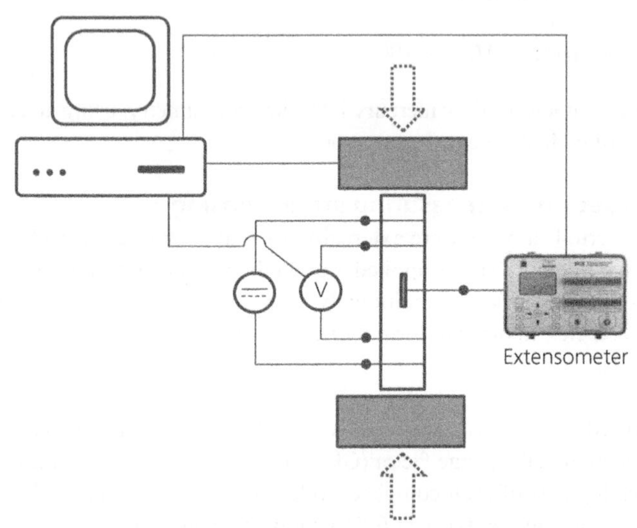

The strain-sensing tests were carried out according to the diagram in Figure 3. For the electrical resistance measures during the tests, an electrical current intensity was set between the outer contacts by means of a DC current source (model Keithley 6220). The voltage was measured between the inner contacts using a digital multimeter (Keithley Model, 2002). Hence resistance may be calculated by applying Ohm's law. Tests were conducted on an electromechanical press model Microtest 10t/2t (Servosis, S.A., Spain). Loads were applied on the smallest faces, that is in the longitudinal direction, and the specimens were perfectly centred on the press to obtain a uniaxial and uniform stress distribution. Strain was permanently monitored with a Vishay P3 extensometer and strain gauges located in the middle point of the lateral sides of the samples, and oriented in the longitudinal direction (the same as the loading direction).

Before and after every test performed on each sample, their masses were controlled. As the mass losses were always below 1% of the initial mass, the water saturation state can be considered constant for all experimental phases; that is, electrical resistance changes are not expected to occur due to water loss (ionic conductivity). Additionally, before any load was applied, the initial electrical resistance was measured. Therefore the initial resistance was controlled in order to calculate the GF after loading the samples.

Results and discussion

Mechanical properties and characterisation tests on CNT cement pastes

Previous compressive strength tests were conducted to determine the loading limits, with the first specimen of each CNT dosage. In order to guarantee an elastic behaviour, the loading conditions during later strain-sensing tests should be below 30% of compressive strength (according to UNE-EN 12390-13:2014). Table 2 shows the specimens' compressive strength mean values. It also includes results of bending strength, ultrasonic pulse velocity, porosity and apparent density for all CNT dosages. Slight variations have been observed for different CNT additions. This trend is consistent with previous research (Camacho-Ballesta *et al.*, 2014). Figure 4 reflects the results obtained from mercury intrusion porosimetry; no meaningful differences are shown for the different dosages analysed.

Table 2 Compressive strength, bending strength, ultrasonic pulse velocity (UPV), porosity and density of different CNT dosage pastes performed at 28 d curing time

CNT dosage: % by cement mass	Compressive strength: N/mm^2	Bending strength: N/mm^2	UPV: km/s	Porosity: %	Density: g/cm^3
0·00	49·2	3·1	3·46	44·2	1·43
0·05	51·2	3·2	3·55	43·5	1·50
0·10	52·4	4·4	3·51	44·2	1·45
0·25	53·0	3·4	3·52	44·7	1·45
0·50	49·0	3·7	3·40	44·7	1·44

Figure 4 Pore diameter distribution for CNT cement pastes resulting from mercury intrusion porosimetry tests

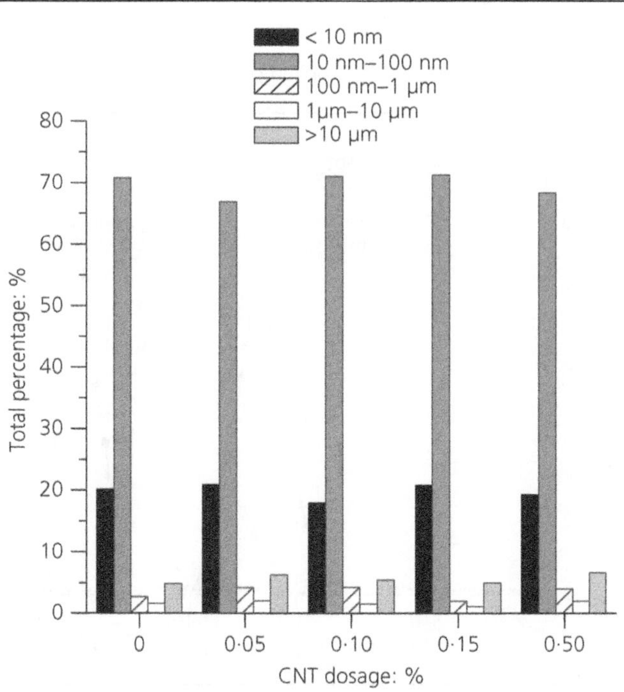

Evaluation of dispersions

When analysing dispersion effectiveness, it was expected that the higher the applied treatment efficacy, the smaller the particle size obtained from the processed light-scattering signal would be. This would mean that CNTs agglomerations are divided into smaller ensembles of CNTs particles, achieving complete dispersion when CNTs can be individually detected. This hypothesis allows the differences found among all applied treatments to be discussed qualitatively. Figure 5 and Figure 6 show the results provided by a light-scattering analyser for applied physical and chemical treatments, respectively. At first sight it seems that physical treatments have a greater dispersant ability than chemical treatments. According to Figure 5, all different mechanical treatments show a particle size smaller than no treatment. Mixer treatment has an indisputable dispersant effect on CNTs by itself, but has no significant contribution when it is combined with subsequent ultrasound treatment, probably owing to a higher energy quantity transferred to CNT conglomerates by ultrasonication technique compared with the mixer treatment. Regarding chemical treatments alone (Figure 6), surfactants would present a certain higher dispersant ability than two polycarboxylates, showing the SDBS the enhanced effect. Nevertheless if a combined dispersant/ultrasonic treatment is applied, differences among them are nullified as a result of the synergistic dispersant effect, as shown in Figure 7 for PC1. Results obtained for the combined treatments with the three other dispersants (PC2, SDBS and TX-100) present an identical behaviour as represented for PC1.

The interesting observation obtained from Figure 7 is that, when combining both chemical and ultrasonication techniques, the particle diameter measured decreases considerably, setting up two well-delineated families. One of them has a mean peak of about 2 μm, which would correspond to a very small bunch of CNTs, the other one is located below 1 μm, which could

Figure 5 Light scattering results obtained for applied physical treatments

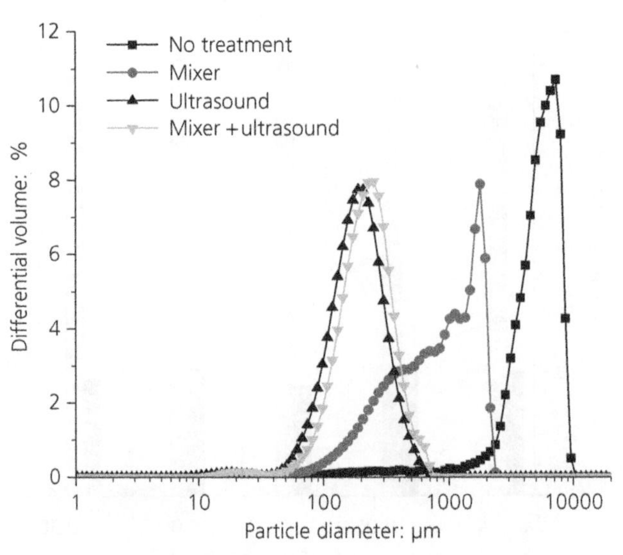

 Light scattering results obtained for applied chemical treatments

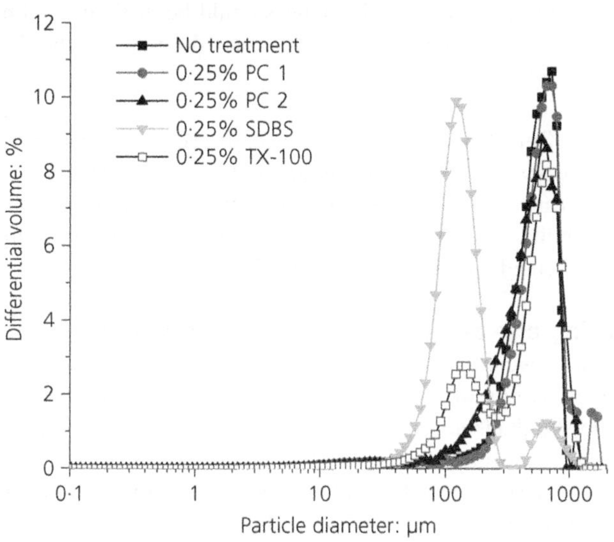

Figure 7 Light scattering results obtained for a combined treatment (physics + different PC1 dosages)

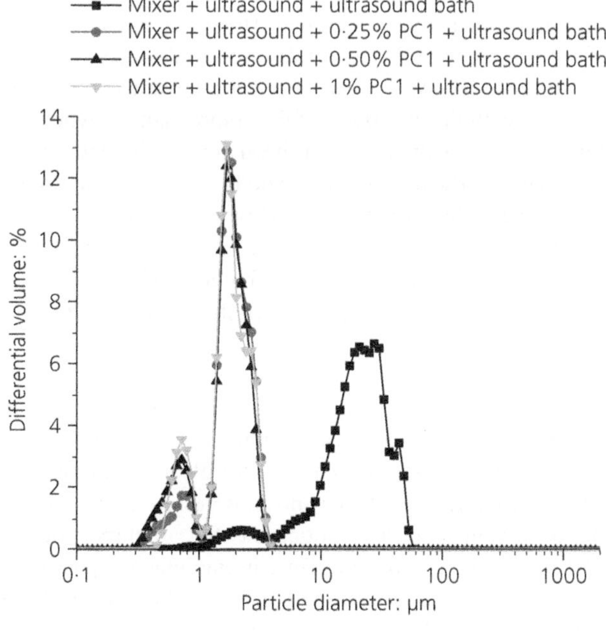

correspond to signals of a single CNT. Therefore, the synergistic effect from physical/chemical combination techniques is found to be the best dispersing method for aqueous CNT suspensions. On the other hand, indiscernible differences could be distinguished between four dispersants, and even more, between different dispersant dosages for the fixed CNT amount (0·05% by cement mass).

Visual observation confirmed the good stability of combined treatment dispersions, not finding any decantation processes at 7, 30 and even 200 d. The 'mixer' treatment showed a partial decantation at 1 d, meanwhile the 'ultrasound' and 'mixer+ultrasound' physical treatments kept a complete stability also at 200 d. Chemical treatments alone, both surfactants and polycarboxylates, presented an immediate decantation process.

Influence of curing age and current intensity on strain-sensing function

The strain-sensing tests were performed on the $40 \times 40 \times 160 \, \text{mm}^3$ specimens. To investigate the piezoresistive behaviour of CNT composites the change in electrical resistance was measured under the simultaneous application of a compressive load. Each regular test consisted of at least four consecutive loading and unloading cycles. The values of load, voltage and unit strain were recorded every 1 s, with the experimental procedure lasting from 120 to 720 s according to the maximum loading and loading rate.

Figure 8 shows the results of strain-sensing tests for the CNT dosages studied (0·05%, 0·10%, 0·25% and 0·50% CNT), at ages of 14 and 28 d. The maximum stress reached during these tests was 3·75 MPa and the loading rate was 400 N/s. Fractional changes in electrical resistance and compressive stress are both represented against time. A relationship between the stress and electrical response could be detected in any of the tests from the earliest tested age of 14 d, for whichever current intensity was applied. However, the strain-sensing performance shows a more sensitive behaviour at the age of 28 d for all dosages, since the reversibility on the measures and the correlation between resistance variation and compressive strain are much better at 28 d than at 14 d, with the exception of the control sample whose resistance changes are not able to be considered reversible. As it has been previously reported (Chen *et al.*, 2014), cement composites with no conductive admixture do not show self-sensing behaviour, or this phenomenon is neither reversible nor repeatable. However, from the lowest CNT addition to the highest, a fractional change in electric resistance has been obtained, which is well correlated with the stress applied to the specimen. This fractional change in resistance shows distinct magnitude for the different CNT dosages, which in turn means different GF for different CNT dosages.

Figure 9 shows the results of strain-sensing tests for 0·50% CNT dosage at a curing age of 28 d. Current intensity was established at three different ranges: 0·1 mA, 1 mA and 10 mA. The maximum stress reached during these tests was 5·625 MPa and the loading rate was 200 N/s. The fractional changes of electrical resistance and compressive stress are both plotted against time. Strain-sensing behaviour is clearly noticeable for all three intensities. It has been observed for all CNT additions that all different currents make it possible to obtain a good reversibility on the measures and establish the correlation between resistance variation and compressive strain, and hence to calculate the corresponding GF. However, sensitivity appears to be enhanced when the current intensity increases. This indicates that, apart from requiring a

Figure 8 Strain-sensing tests for different CNT dosages 0·05%, 0·10%, 0·25% and 0·50% CNT at curing ages of 14 and 28 d. Fractional change on electrical resistance, $\Delta R/R_0$ and compressive stress are both represented against time. All tests were run at a constant loading rate of 400 N/s and a current intensity of 1 mA: (a) 0·05% CNT; (b) 0·10% CNT; (c) 0·25 CNT; (d) 0·5% CNT

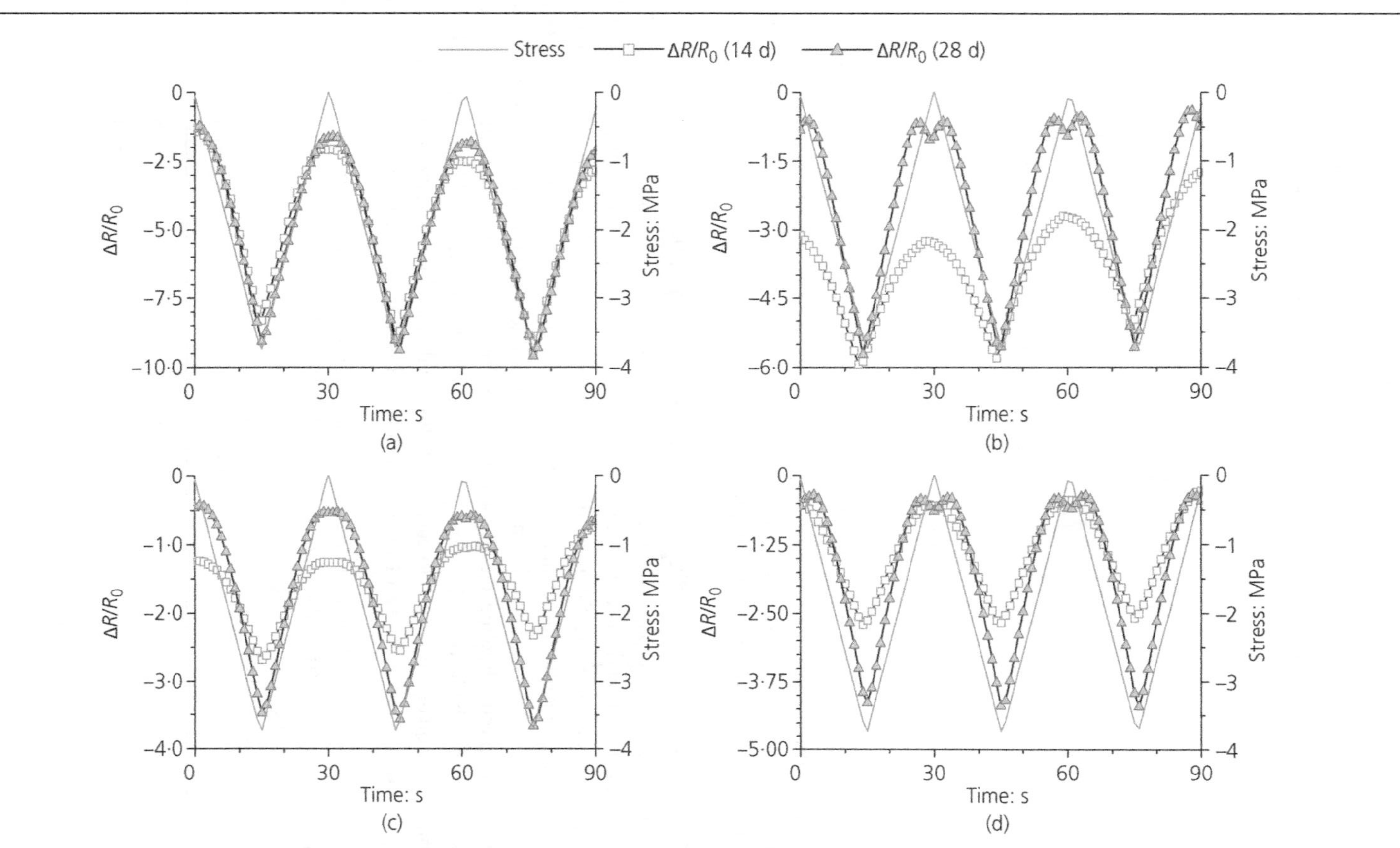

Figure 9 Fractional change in electrical resistance ($\Delta R/R_0$) and longitudinal unit strain (µε) plotted against time for 0·50% CNT cement pastes strain-sensing tests, at 28 d, for 0·1 mA, 1·0 mA and 10·0 mA current intensities, at a maximum load of 9 kN (5·625 MPa) and a loading rate of 200 N/s

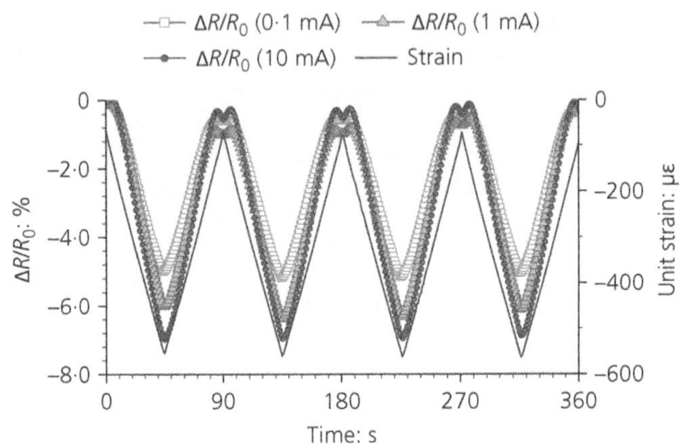

certain current density for the piezoresistive behaviour to be visible (Galao *et al.*, 2014), there is an optimal current density which enables the best correlation levels between electric and mechanical parameters to be obtained.

Effect of loading conditions on strain-sensing function

In the elastic range pastes behave according to $\sigma = E\varepsilon$, where σ is stress (MPa), E is elastic modulus (MPa) and ε is unit strain. For a specific stress, the registered unit strain will depend on the elastic modulus. So given the influence of ε for GF calculation, it is also convenient to take elastic modulus values into account. Table 3 shows the mean elastic modulus (E) obtained at the age of 28 d for a maximum load of 9 kN and loading rate of 100 N/s. Slight differences were observed, showing the highest E value for the control sample and the lowest elastic modulus for the lowest CNT dosage. These slight differences could be explained by different plasticiser quantities added to each CNT dosage that enable the same workability but also can generate a small increase of matrix stiffness. However, given the similarity of E values for

Table 3 Mean elastic modulus (E) obtained for fixed load conditions: maximum load of 9 kN and loading rate of 100 N/s. Average value, standard deviation (SD) and relative SD obtained for three measures performed at 28 d

% CNT (by cement mass)	Elastic modulus: N/mm²	SD: N/mm²	RSD: %
0·00	12 422	263	2·1
0·05	13 101	235	1·8
0·10	13 574	52	0·4
0·25	10 989	396	3·6
0·50	10 705	318	3·0

pastes incorporating CNT, a similar mechanical behaviour can be expected for these samples when they are subjected to stress during strain-sensing tests. Regarding the strain distribution in specimens, all samples are expected to experience the same uniaxial tensile stress since they are geometrically equal. Research is presently underway within a structural reinforced element, which also includes other sorts of sensors that are different in shape.

The loading conditions refer to the combination of different maximum loadings (3, 6, 9 kN) and loading rates (50, 100, 200 and 400 N/s). After carrying out the nine combinations established in the experimental programme, the pattern plotted in Figure 10 was seen repeatedly. It shows the gauge factor of the 0·50% CNT sample for each maximum cycle loading, considering unloading cycles as negative loading rates and loading ones as positive rates. If a certain load level is reached the material's response is independent of the loading rate used (similar conclusions could be drawn for different moving loads' velocities reported in previous research (Baeza *et al.*, 2013a)). This topic is especially important if traffic monitoring applications are desired, because if the materials' response is independent of the loading rate, it will be independent of the vehicle's passing velocity. However, the stress applied to the samples does affect the sensitivity of the CNT cement-based pastes. At the lowest loading level (3 kN) a loss in sensitivity has been registered for all different pastes; that is, composites must experience a minimum strain in order to show improved strain-sensing properties. Besides, once this sensitivity threshold is surpassed the gauge factors continue depending on the stress, but obtain more similar values between the two highest values (6 and 9 kN).

Figure 10 Gauge factor of 0·50% CNT paste at 28 d against loading rate (negative values correspond to unloading cycles) for three different maximum axial loads (the stress levels applied were 1·875, 3·75 and 5·625 MPa, for 3, 6 and 9 kN, respectively)

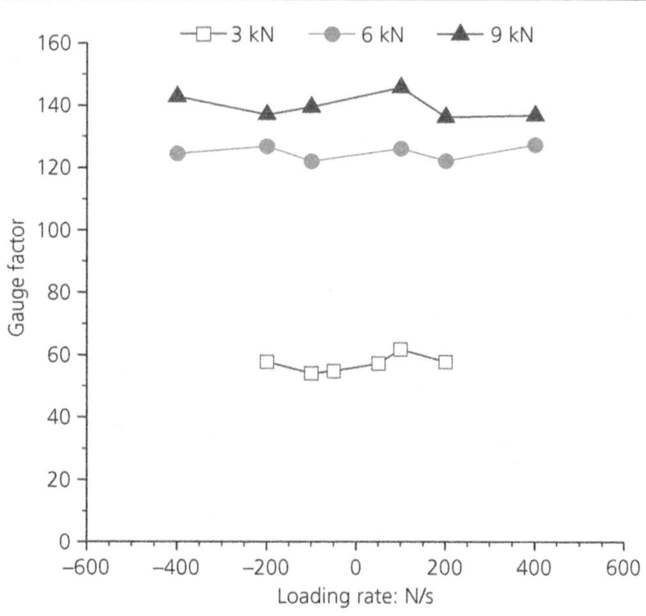

Figure 11 includes the mean GF values calculated for specimens with CNT additions, for the maximum load applied of 9 kN, that is the maximum load that had been found to show the enhanced sensitivity. Results are shown for the loading rate of 400 N/s, but for all three different loading rates the obtained behaviour pattern is analogous, according to the non-influence of the loading rate mentioned above. The R^2 Pearson's coefficient is also plotted, obtained as the average value for both loading and unloading GF linear regressions. Three interesting conclusions can be drawn from these results.

(a) There were no major differences between loading and unloading cycles, thus the analysis of the mean value of all cycles would be accurate enough.
(b) R^2 Pearson's coefficient shows a pattern similar to GF values, that is, the higher GF, the better the R^2 coefficient.
(c) The addition of CNT to the cement paste achieves a gauge factor up to 240, which is reached for the lowest CNT dosage. The most likely explanation for this observation is a better dispersion for 0·05% CNT into the matrix. That would mean the same dispersion method applied for all samples is not equally effective for different CNT dosages.

The curves in Figure 12 represent the electrical data (fractional change of resistance) against the mechanical data (strain) when the measures of the same single test (maximum loading 9 kN and loading rate 400 N/s) are represented for different CNT dosages. Linear regressions for the test appear defined for each dosage by their equation, where the slope is GF

Figure 11 Gauge factor with standard deviation and its R^2 Pearson's coefficient for different CNT dosages at the age of 28 d. Maximum loading 9 kN and loading rate 400 N/s. Current intensity established at 1 mA

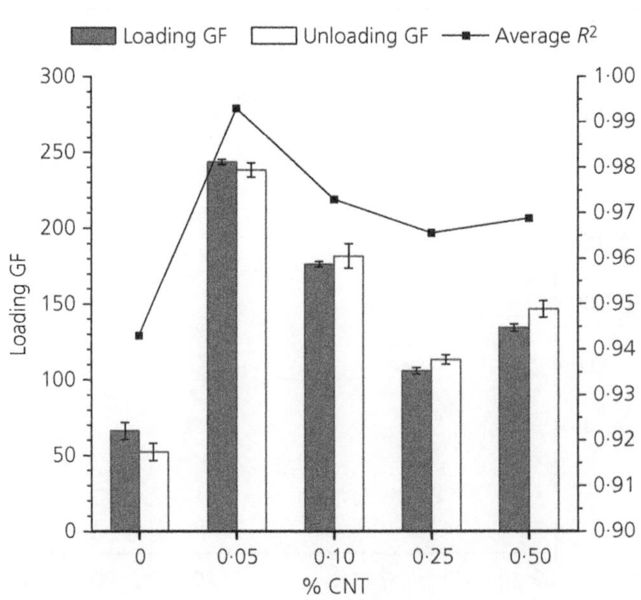

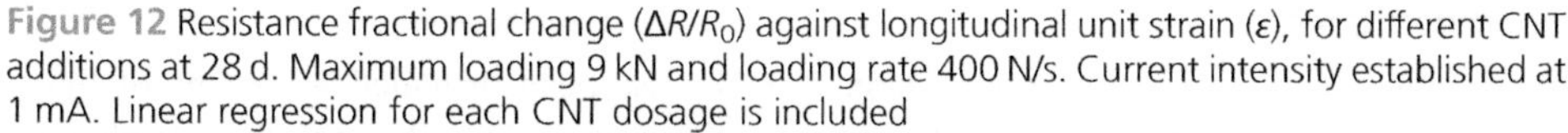

Figure 12 Resistance fractional change ($\Delta R/R_0$) against longitudinal unit strain (ε), for different CNT additions at 28 d. Maximum loading 9 kN and loading rate 400 N/s. Current intensity established at 1 mA. Linear regression for each CNT dosage is included

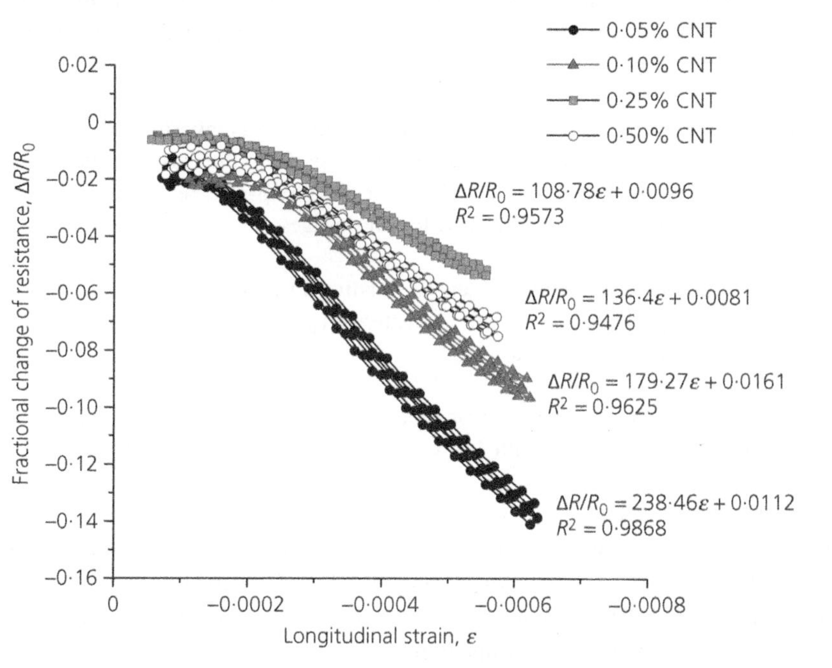

according to Equation 1, and their R^2 Pearson's coefficient. Here, the behaviour pattern described for Figure 11 is also identifiable; that is, the highest GF is equally shown by the lowest CNT dosage and it also presents the highest R^2. At this point the suitability of the GF as a characteristic parameter could be assessed. When analysing the sensing phenomena at the lowest strains, there seems to be a low sensitivity area below 200 µε, which would explain the lower GF values for very low stresses. However, when the 200 µε strain value was exceeded, a perfectly linear resistivity change–strain relationship was found. Therefore, according to Equation 1, the GF description would be only available for this second case.

In any case, the non-linear singularity at low strains had a small effect on all topics discussed above, as the linear regression curve calculated for every test had high average R^2 coefficients. Thus GF can actually be used for strain-sensing purposes as proposed by several authors (Baeza *et al.*, 2011; Chung, 2002) (even if third-degree polynomial or sigmoid functions could make better regressions), but certain sensitivity issues should be considered for real applications.

Lastly, as far as durability is concerned, as in previous research (Camacho-Ballesta *et al.*, 2014) it seems to be appropriate to do research into the effect of aggressive conditions, such as carbonation and contamination by chloride ions, in the strain-sensing function. New experimental tests are already planned to that effect.

Conclusions

An insight into the problem of dispersion of CNT has been provided by means of light-scattering particle size analysis. This technique made it possible to discern which treatments presented the best dispersant ability, that is, the combined physical/chemical treatment (ultrasound + dispersant), but did not show any difference among the distinct dispersants evaluated.

The piezoresistive sensitivity of cement-based CNT sensors was investigated at the ages of 14 and 28 d. Strain-sensing properties were seen in all tests from the earliest tested age of 14 d, whichever current intensity was applied. However, the strain-sensing performance showed a more sensitive behaviour at the age of 28 d for all dosages.

As far as loading conditions were concerned, loading rate did not affect the strain-sensing response of the composites, although the sensitivity (gauge factor) was increased with the maximum compressive load applied.

The best performance as strain sensor was obtained for the 0·05% CNT composite, reaching values of gauge factor up to 240 with R^2 Pearson's coefficient above 0·99.

Acknowledgements

The authors would like to acknowledge the Spanish Ministry of Science and Innovation (Ref: Mat 2009-10866) and Generalitat Valenciana (PROMETEO/2013/035) for their financial support of this research. The authors would also like to thank Bayer MaterialScience, S.A. for the supply of CNTs used in this investigation.

REFERENCES

AENOR (Spanish Association for Standardisation and Certification) (2005) UNE-EN 196-1:2005. Methods of testing cement – Part 1: Determination of strength. Spanish Association for Standardisation and Certification, Madrid, Spain.

AENOR (2009) UNE-EN 196-3:2005+A1:2009. Methods of testing cement – Part 3: Determination of setting times and soundness. Spanish Association for Standardisation and Certification, Madrid, Spain.

AENOR (2014) UNE-EN 12390-13:2014. Testing hardened concrete – Part 13: Determination of secant modulus of elasticity in compression. Spanish Association for Standardisation and Certification, Madrid, Spain.

Azhari F and Banthia N (2012) Cement-based sensors with carbon fibers and carbon nanotubes for piezoresistive sensing. *Cement and Concrete Composites* **34(7)**: 866–873.

Baeza FJ, Galao O, Zornoza E and Garces P (2013a) Effect of aspect ratio on strain sensing capacity of carbon fiber reinforced cement composites. *Materials and Design* **51(October)**: 1085–1094.

Baeza FJ, Galao O, Zornoza E and Garces P (2013b) Multifunctional cement composites strain and damage sensors applied on reinforced concrete (RC) structural elements. *Materials* **6(3)**: 841–855.

Baeza FJ, Zornoza E, Andion LG, Ivorra S and Garces P (2011) Variables affecting strain sensing function in cementitious composites with carbon fibers. *Computers and Concrete* **8(2)**: 229–241.

Bortz DR, Merino C and Martin-Gullon I (2011) Carbon nanofibers enhance the fracture toughness and fatigue performance of a structural epoxy system. *Composites Science and Technology* **71(1)**: 31–38.

Camacho-Ballesta MC, Galao O, Baeza FJ, Zornoza E and Garces P (2014) Mechanical properties and durability of CNT cement composites. *Materials* **7(3)**: 1640–1651.

Catala G, Ramos-Fernandez EV, Zornoza E, Andion LG and Garces P (2011) Influence of the oxidation process of carbon material on the mechanical properties of cement mortars. *Journal of Materials in Civil Engineering* **23(3)**: 321–329.

Chen SJ, Collins FG, Macleod AJN *et al.* (2011) Carbon nanotube–cement composites: a retrospect. *The IES Journal Part A: Civil and Structural Engineering* **4(4)**: 254–265.

Chen SJ, Zou B, Collins F *et al.* (2014) Predicting the influence of ultrasonication energy on the reinforcing efficiency of carbon nanotubes. *Carbon* **77(2014)**: 1–10.

Chung DDL (2002) Piezoresistive cement-based materials for strain sensing. *Journal of Intelligent Material Systems and Structures* **13(9)**: 599–609.

Chung DDL (2012) Carbon materials for structural self-sensing, electromagnetic shielding and thermal interfacing. *Carbon* **50(9)**: 3342–3353.

Coleman JN, Khan U, Blau WJ, Gun'Ko YK (2006) Small but strong: a review of the mechanical properties of carbon nanotube-polymer composites. *Carbon* **44(9)**: 1624–1652.

Collins F, Lambert J and Duan WH (2012) The influences of admixtures on the dispersion, workability, and strength of carbon nanotube-OPC paste mixtures. *Cement and Concrete Composites* **34(2)**: 201–207.

Galao O, Zornoza E, Baeza FJ, Bernabeu A and Garces P (2012) Effect of carbon nanofiber addition in the mechanical properties and durability of cementitious materials. *Materiales de Construccion* **62(307)**: 343–357.

Galao O, Baeza FJ, Zornoza E and Garces P (2014) Strain and damage sensing properties on multifunctional cement composites with CNF admixture. *Cement and Concrete Composites* **46(2014)**: 90–98.

Han B, Yu X and Kwon E (2009) A self-sensing carbon nanotube/cement composite for traffic monitoring. *Nanotechnology* **20(44)**: 445501.

Han B, Yu X, Kwon E and Ou J (2012) Effects of CNT concentration level and water/cement ratio on the piezoresistivity of CNT/cement composites. *Journal of Composite Materials* **46(1)**: 19–25.

Kim HK, Nam IW and Lee HK (2014) Enhanced effect of carbon nanotube on mechanical and electrical properties of cement composites by incorporation of silica fume. *Composite Structures* **107(January)**: 60–69.

Konsta-Gdoutos MS and Aza CA (2014) Self sensing carbon nanotube (CNT) and nanofiber (CNF) cementitious composites for real time damage assessment in smart structures. *Cement and Concrete Composites* **53(October)**: 162–169.

Li GY, Wang PM and Zhao X (2007) Pressure-sensitive properties and microstructure of carbon nanotube reinforced cement composites. *Cement and Concrete Composites* **29(5)**: 377–382.

Luo J, Duan Z and Li H (2009) The influence of surfactants on the processing of multi-walled carbon nanotubes in reinforced cement matrix composites. *Physica Status Solidi A* **206(12)**: 2783–2790.

Luo J, Duan Z, Zhao T and Li Q (2011) Effect of compressive strain on electrical resistivity of carbon nanotube cement-based composites. *Key Engineering Materials* **483**: 579–583.

Parveen S, Rana S and Fangueiro R (2013) A review on nanomaterial dispersion, microstructure, and mechanical properties of carbon nanotube and nanofiber reinforced cementitious composites. *Journal of Nanomaterials* **2013(2013)**: 710175.

Wen S and Chung DDL (2006) Model of piezoresistivity in carbon fiber cement. *Cement and Concrete Research* **36(10)**: 1879–1885.

Yu X and Kwon E (2009) A carbon nanotube/cement composite with piezoresistive properties. *Smart Materials and Structures* **18(5)**: 055010.

Dhir and Paine
ISBN 978-0-7277-6457-7
https://doi.org/10.1680/icetsc.64577.225
ICE Publishing: All rights reserved

Chapter 14

Novel concrete-temperature monitoring method by using embedded passive wireless sensor

Hong Chen
School of Transportation and Logistics, East China Jiaotong University, Nanchang, China

Shuangxi Zhou
School of Civil Engineering and Architecture, East China Jiaotong University, Nanchang, China (corresponding author: 605460326@qq.com)

Fangming Deng
School of Electrical and Electronic Engineering, East China Jiaotong University, Nanchang, China

Jiayu Zou
School of Civil Engineering and Architecture, East China Jiaotong University, Nanchang, China

Xiang Wu
School of Electrical and Electronic Engineering, East China Jiaotong University, Nanchang, China

Zhihui Fu
School of Electrical and Electronic Engineering, East China Jiaotong University, Nanchang, China

This paper presents a novel concrete-temperature monitoring method using an embedded passive wireless sensor. After introducing the control method of concrete-temperature monitoring, this work proposes an embedded passive wireless sensor based on radio frequency identification (RFID) technology. It includes a microstrip antenna specially designed for embedded monitoring due to its round plane architecture. The proposed wireless sensor consists of three parts: an energy management section, a digital section and an RFID section. Laboratory experimentation and on-site testing demonstrate that, despite the high losses of concrete, the proposed sensor tag provides reliable communication performance in passive mode. The maximum reader–tag communication distance is 7 m at 915 MHz, and the measured results are highly consistent with the results measured by thermocouple. To the knowledge of the authors, this work is the first to use the RFID sensor tag to measure concrete temperature.

Notation

c	specific heat capacity
c_c	concrete heat capacity
c_w	pipe heat capacity

d	communication distance
E_r	effective isotropic radiation power of a reader
f	frequency
G_a	tag antenna gain
h_y	degree of hydration
m	external humidity
p	location
P_t	operating power of an RFID tag
Q	power density of the internal heat source per volume
q_w	water flow through a pipe
t	time
T	temperature
$T_{w\text{-}in}$	water inflow of pipe
$T_{w\text{-}out}$	water discharge of pipe
V_c	concrete volume
ε	permittivity
η_r	RF-to-DC power conversion efficiency of the rectifier
λ	wavelength of the electromagnetic wave
ρ_c	concrete density
ρ_w	pipe density
σ	conductivity
τ	time variable

Introduction

Concrete material properties (i.e. concrete strength, modulus of elasticity, creep and shrinkage) change with time, and these properties are significantly influenced by the heat of hydration and moisture content in the concrete during the early stages (Wang *et al.*, 2015). Thus, in many areas of civil engineering, especially in concrete maturity, information on the temperature of concrete is extremely important (Bicanic and Zhang, 2006). The majority of civil structures, such as tunnels, bridges and buildings, are transported to the construction site after casting and curing, and are built by assembling precast concrete blocks (segments). Curing temperature is one of the parameters considered as critical factors in the progress of cement hydration, influencing the stability and transformation of hydrates and strength development (Tang and Lo, 2009). The result of this process is an increased strength and decreased permeability. Uncontrolled high temperature development during the setting and curing process can cause a number of detrimental effects. For example, fast moisture diffusion resulting from a high temperature during curing may lead to higher shrinkage stresses and lower long-term concrete strength (Fu *et al.*, 2005). Such a temperature can also promote the deterioration processes of concrete structures. However, making good use of high temperatures during the maturation process can bring many benefits. Especially by heating the ashlar of fresh concrete in a controlled way, concrete can develop a good mechanical strength in a few hours rather than after the conventional 28 d, resulting in an increase in productivity. Therefore, the measurement of temperature is significant for concrete.

Traditional measuring methods always employ thermocouples (Lothenbach *et al.*, 2007; Yan *et al.*, 2013; Zhang *et al.*, 2013) and measure the temperature at the centre of the concrete. This method has a relatively high cost, and the presence of cables that must be removed after the measurement process prevents this technology from being widely applied. Hence, the maturation quality of concrete is measured only for a few samples. A passive and wireless device has the potential to solve such problems and to make the pervasive displacement of temperature sensors a reality. Radio frequency identification (RFID), as a wireless automatic identification technology, is widely applied, for example in traffic management, logistics, medicine management and food production (Lei *et al.*, 2013). Passive RFID tags offer several advantages such as battery-less operation, wireless communication capability, high flexibility, low cost and fast deployment, which have all resulted in their extensive use in commercial applications (Schindler *et al.*, 2006). The passive RFID tag uses the wireless energy from the RFID reader as its power supply. Recently, with the rapid development of the Internet of Things and sensor technology, research on adding sensing functionality to RFID tags has become a hot topic (Deng *et al.*, 2015a, 2015b; Zuo *et al.*, 2012). These smart RFID sensing tags not only extend the fields of application of RFID, but also contribute to reducing the cost of RFID system fabrication.

This paper presents a novel wireless sensor for concrete temperature measurement. This sensor is based on ultra-high frequency (UHF) RFID technology. Compared with traditional temperature measurement, the proposed sensor can measure the temperature in a wireless way without any damage to the concrete. Furthermore, the proposed passive RFID sensor tag can work without a battery for low-cost application. The rest of the paper is organised as follows. The section entitled 'Intelligent temperature control: method and principle' introduces the theory of intelligent temperature control. The section 'Wireless sensor tag architecture' presents the overall architecture of the proposed sensor tag and then analyses this design in detail, while the section 'Experimental characterisation' illustrates the measurement results and compares it with the results of traditional measurement. The paper ends with a 'Conclusion'.

Intelligent temperature control: method and principle

Temperature is one of the main parameters in concrete monitoring. In order to avoid concrete cracking during the concrete cooling process, three basic principles need to be considered: maximum temperature control, the rate of temperature change control and abnormal temperature control (LéGer and Leclerc, 2007; Zhang, 2007). Firstly, a large temperature difference can lead to concrete cracking and hence the maximum internal temperature should be different according to the ambient temperature. Secondly, rapid temperature change may cause concrete cracking, and thus the control of temperature change should be considered seriously. Lastly, a combination of higher summer temperatures and lower winter snowfall may also cause concrete cracking. In these situations, a control system should be adopted to decrease the temperature stress.

Regarding the internal heat source, a basic equation of three-dimensional heat conduction can be obtained by using Fourier's Law (Schindler *et al.*, 2006)

$$\frac{\Delta T}{\Delta \tau} = \left(\frac{\Delta^2 T}{\Delta x^2} + \frac{\Delta^2 T}{\Delta y^2} + \frac{\Delta^2 T}{\Delta z^2} \right) + \frac{Q}{\rho c} \tag{1}$$

where T represents temperature, τ the time variable, c specific heat capacity and Q the power density of the internal heat source per volume. Considering the temperature field as a steady field, we can get the following equation

$$\frac{\Delta T}{\Delta \tau} = \frac{\Delta \theta(\tau)}{\Delta \tau} + \frac{\rho_w c_w}{\rho_c c_c} \frac{q_w [T_{w\text{-in}}(\tau) - T_{w\text{-out}}(\tau)]}{V_c} \tag{2}$$

where $\rho_{\rm w}$, $c_{\rm w}$ and $q_{\rm w}$ represent the density, specific heat capacity and water flow through a pipe, respectively; $\rho_{\rm c}$, $c_{\rm c}$ and $V_{\rm c}$ represent the density, specific heat capacity and volume of concrete, respectively; and $T_{\rm w-in}$ and $T_{\rm w-out}$ represent water inflow and discharge through a pipe, respectively. According to the target temperature, we can get the theoretical water flow in Equation 3

$$q_{\rm w}(\tau) = \frac{[T(\tau+\Delta\tau) - T(\tau)]/\Delta\tau - \theta'(\tau)}{\rho_{\rm w}(\tau)c_{\rm w}(\tau)[T_{\rm w-in}(\tau) - T_{\rm w-out}(\tau)]/\rho_{\rm c}c_{\rm c}V_{\rm C}} \tag{3}$$

Therefore, a basic flow of temperature control, proposed in Figure 1, can be acquired from the above equations.

Figure 1 Flow diagram of the concrete temperature control

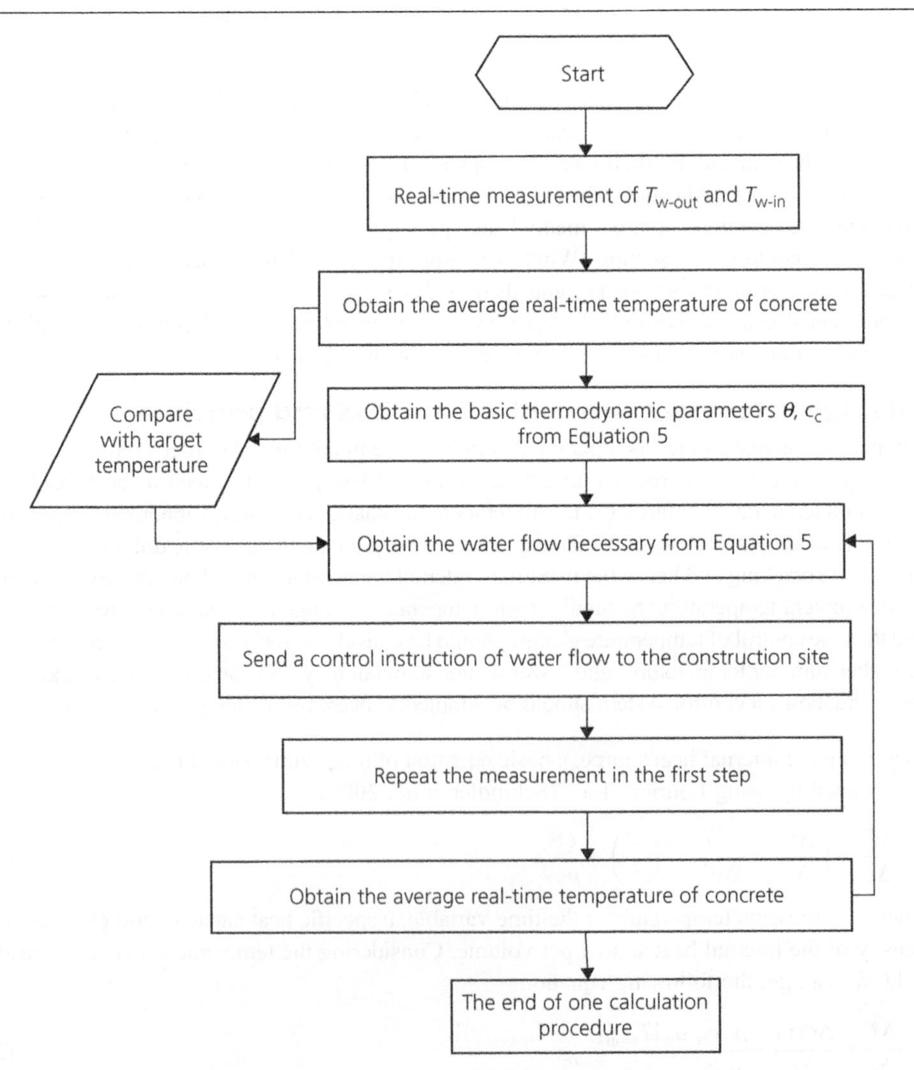

Wireless sensor tag architecture

Concrete is a mixture of cement, water and aggregated composites, and the electromagnetic field is highly affected by the dielectric properties of the medium. Since concrete is a lossy material, its influence on RFID communication needs to be seriously taken into account. Concrete dielectric properties, such as the permittivity $\varepsilon(T; m; p; f; \text{w/c}; c)$ and the conductivity $\sigma(T; m; p; f; \text{w/c}; c)$, are generally related to the temperature T, the external humidity m, the frequency f, the time t spent since casting, the location p, the water/cement ratio (w/c) and the type of cement and its specific heat capacity c. By fitting data from previous references (Kim *et al.*, 2011; Klysz *et al.*, 2008), it is possible to estimate the trend of conductivity and permittivity, while the degree of hydration h_y changes at UHFs during the concrete maturation process (shown in Figure 2), which will be applied in the following numerical analysis. According to Figure 2, while the conductivity decreases proportionally to the degree of hydration, we can find that the trend of the permittivity is not monotonic and reaches its peak at about one-fifth of the way through the maturation process.

Because the microstrip antenna includes a ground plane architecture, it has a natural anti-metal advantage compared to the dipole antenna. Therefore, the microstrip antenna is much more suitable for embedded concrete health monitoring than the dipole antenna. The length of a rectangular microstrip patch antenna is approximately equal to half of its electrical length,

Figure 2 Performance analysis of concrete at 915 MHz; (a) permittivity of concrete vs degree of hydration; (b) conductivity vs degree of hydration

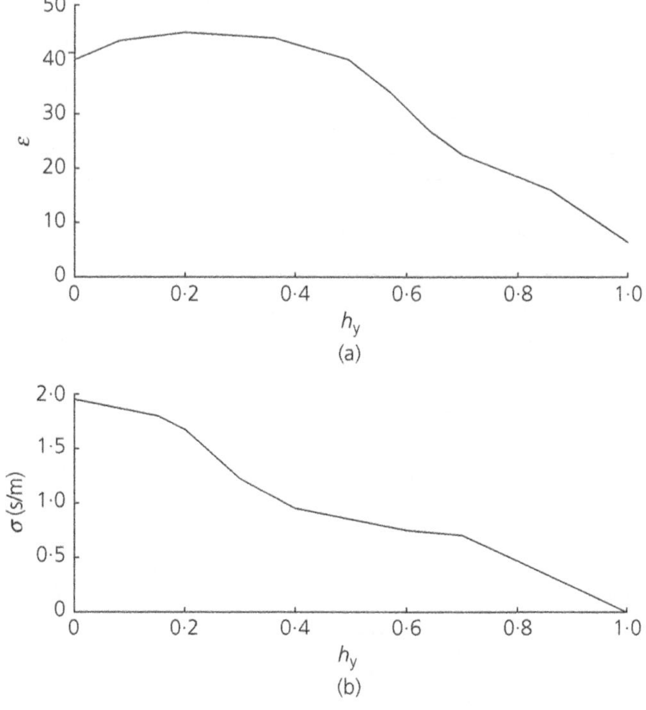

which is not suitable for the miniaturisation of tags. As shown in Figure 3, we presented a miniaturised microstrip antenna using the embedded short stub and U-type slot. The impedance of the chip was 19-j172 Ω at 915 MHz. The copper traces, with a thickness of 0·035 mm, were printed on an FR4 substrate, whose permittivity was 4·4 and thickness was 2 mm. The current is strongest in the short-circuited stub of this antenna, so the impedance of the antenna can be adjusted easily by adjusting the value of L_{F1} and L_{F2} (see Figure 3). The short stub embedded in the radiation patch internally can lead to the reduction of the antenna volume. Furthermore, because the current can flow around the slot, this results in increasing the length of the current path and furthers miniaturisation.

Figure 4 shows the architecture of the proposed wireless sensor tag. The sensor tag is based on the EPC Generation-2 UHF communication protocol and operates in passive mode

Figure 3 Layout of the proposed T-type antenna

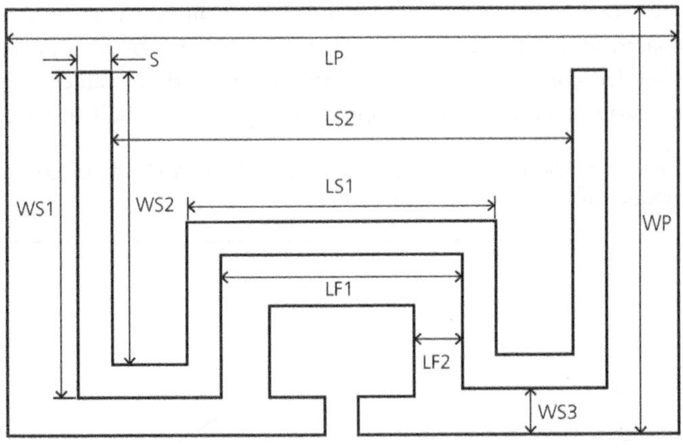

Figure 4 Architecture of the proposed wireless sensor tag

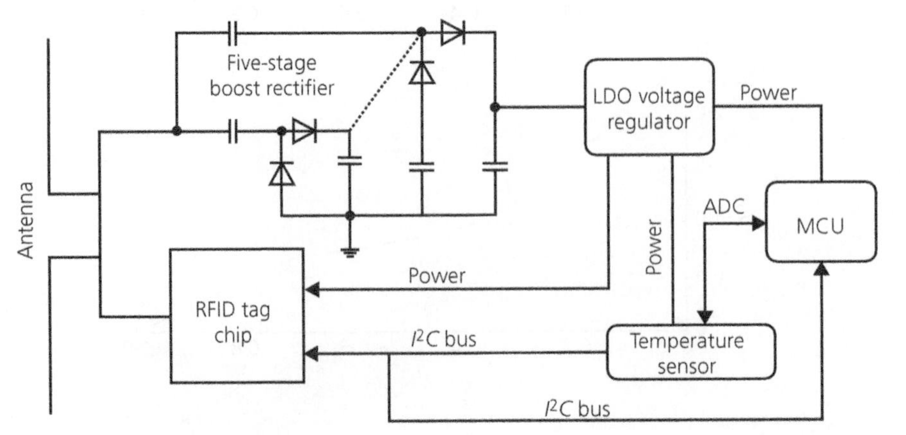

(Catarinucci *et al.*, 2011, 2012; De Donno *et al.*, 2014). The whole sensor tag consists of three sections: the energy management section, the digital section, and the RFID section. A five-stage boost rectifier along with a low-dropout (LDO) regulator make up the energy management section. The digital section includes a MicroController Unit (MCU), temperature sensor and I^2C interface. Except for the traditional air interface, the memory of the proposed temperature sensor tag can also be accessed by the I^2C interface. In this way, the sensor data can be transferred to the sensor tag memory by the I^2C bus and then acquired by a standard RFID reader. The RFID section is composed of a standard Gen2 RFID tag chip and the proposed microstrip antenna.

The theoretical practicable operating power of an RFID tag P_t is calculated from the Friis transmission equation (Wei *et al.*, 2011)

$$P_t = E_r G_a \eta_r \left(\frac{\lambda}{4\pi d} \right)^2 \tag{4}$$

where E_r is the effective isotropic radiation power of a reader, G_a is the tag antenna gain, η_r is the RF-to-DC power conversion efficiency of the rectifier, λ is the wavelength of the electromagnetic wave and d is the communication distance. According to Equation 4, the communication distance d can be described as follows

$$d = \frac{\lambda}{4\pi} \sqrt{\frac{E_r G_a \eta_r}{P_t}} \tag{5}$$

From this equation, it can be concluded that, in order to achieve a longer communication distance, a lower power of the tag and a higher η_r of the rectifier are critical for the UHF RFID tag design because E_r is limited by regional regulations (4 W is the maximum transmitted power) and G_a is roughly determined by the allowable antenna area (1·64 for the $\lambda/2$ dipole antenna).

Experimental characterisation

Figure 5(a) is a photograph of our RFID sensor tag. The RFID sensor tag was designed on FR4 substrate using discrete components. This tag chip was matched with the packaged microstrip antenna on FR4 substrate for the measurement of temperature in the concrete. The sensor tag was first calibrated and tested in the laboratory, as shown in Figure 5(b). The VISN-R1200 is a special RFID tester from the VI Service Network that can process, analyse and display the measurement results simultaneously. An anechoic box was used to provide an enclosed electromagnetic measuring environment. The temperature performance of the sensor tag was calibrated in a climate chamber (VCL 4003).

To verify the RFID communication of the proposed sensor tag, the testing parameters are chosen as follows: the operating frequency is 915 MHz, the distances between the two antennas and the sensor tag are both 0·5 m, the transmitting power is 1 W. The measured communication flow is shown in Figure 6. The RFID tester first sends a select instruction of inventory sequence order. After waiting for 5–6 Tari, the RFID tester then sends a query

Figure 5 Proposed RFID sensor tag: (a) photograph of the proposed tag; (b) measuring environment

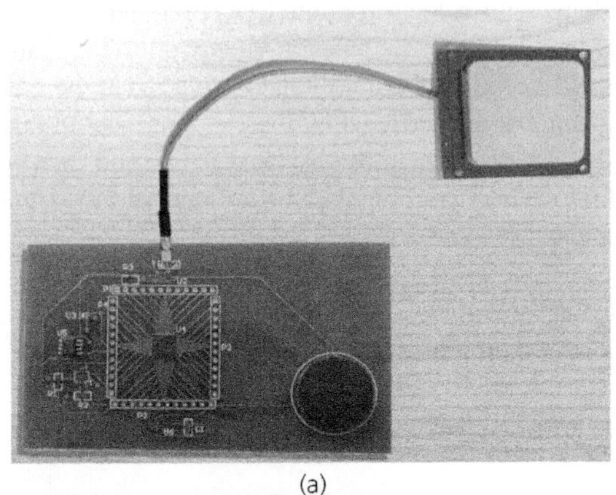

(a)

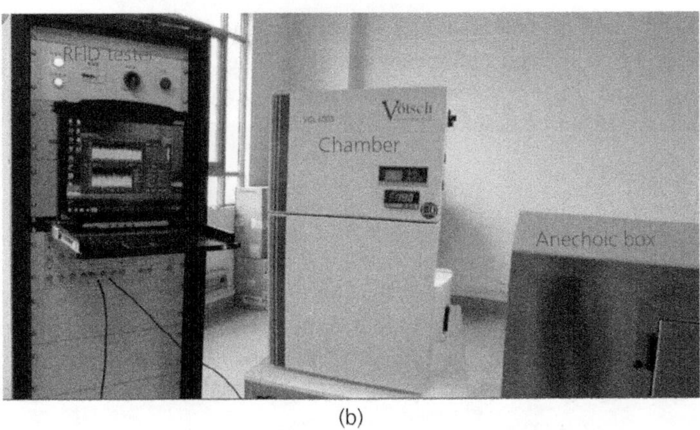

(b)

Figure 6 Measured communication flow (UII, user information interaction)

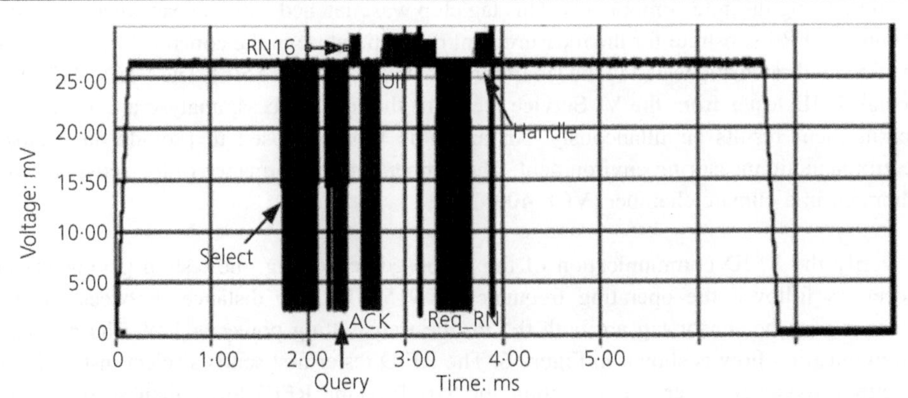

instruction and the tag responds back with RN16. The tester then sends an ACK instruction to acquire the ID information of the tag including the sensor data. After that, the tester sends a ReqRN instruction to obtain the handle response.

The maximum reader–tag communication distance is definitely one of the most important metrics for characterising the performance of UHF RFID tags (De Donno *et al.*, 2013a, 2013b; Vyas *et al.*, 2011). According to Equation 5, $E_{rmax} = 4$ W is the maximum reader E_r allowed by the corresponding regulations. In order to experimentally measure the communication distance, a commercial Gen2 reader, operating in the 750–950 MHz frequency band with equivalent isotropically radiated power set to $E_{rmax} = 1$ W, was used to interrogate the RFID tag at different distances. As shown in Figure 7(a), the sensor tag was immersed in the fresh concrete with its antenna outside the concrete. Both the tag and reader are oriented in the maximum-gain direction. The reader–tag distance was increased in steps of 0·5 m, and for each measurement point the reader was instructed to perform 1000 attempts to read the sensor data. If the success

Figure 7 Maximum communication distance measurement: (a) measuring environment; (b) measuring results

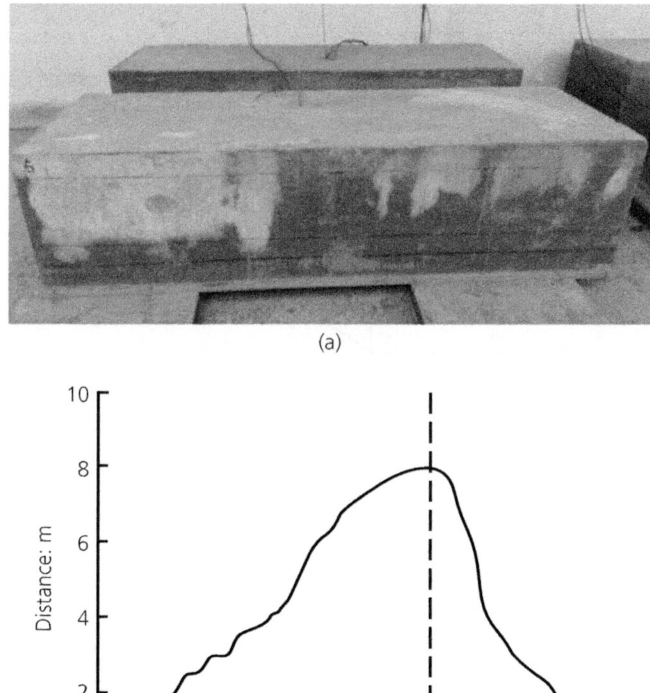

(a)

(b)

ratio remains above 80%, the distance is regarded as a reliable communication distance. The test result is shown in Figure 7(b), and it can be concluded that the maximum reader–tag communication distance is approximately 7 m at 915 MHz.

Finally, the temperature performances of the concrete during the maturation process were tested by the proposed RFID sensor. As shown in Figure 8(a), the tests were repeated when the RFID sensor was immersed at depths of 15 cm (A point), 10 cm (B point) and 5 cm (C point). respectively. Three thermocouples were, moreover, placed in the same position to provide a reference measurement. During the following 3 d after casting, the concrete specimens were exposed to the ambient temperature and the temperature data was recorded every 30 min. Figure 8(b) compares the temperature data measured by the RFID sensor and thermocouple, which almost follow the same profile. The maximum temperature rise of three curves is 25°C and the maximum difference between the temperatures of the three curves at the same time is lower than 25°C, which meets the design requirements. About 15 h after casting, the temperatures of the three test points all reach their peak, proving that the hydration heat plays a vital role in the variation of the internal temperature of concrete. Due to the different depths of the three test points, the rates of heat accumulation and dissipation are different, resulting in different peak values and different drop rates for the measured temperature.

Figure 8 Measured internal temperature of concrete: (a) measuring site; (b) measured performances

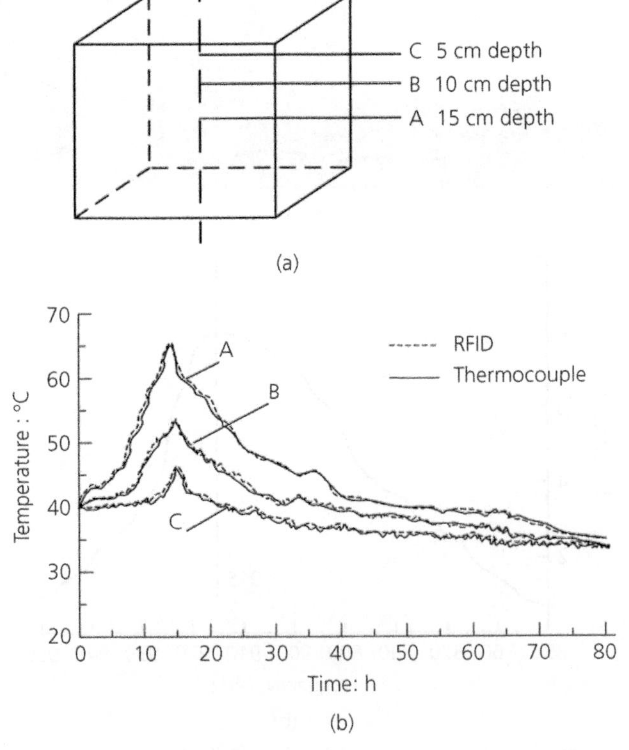

Conclusion

Temperature monitoring plays an important role in concrete measurement, and thus this work presents a wireless temperature sensor tag based on the EPC Gen-2 UHF communication protocol. Considering the high losses of electromagnetic waves in concrete, a T-type antenna is proposed to ensure that the sensor tag can work inside the concrete. The communication flow of the proposed sensor tag was tested in a laboratory, while the maximum operating distance and temperature measurement were tested on a construction site. The obtained communication distance is 7 m at 915 MHz, and the data measured by the RFID sensor and thermocouple follow almost the same profile. The whole experiment demonstrates that the proposed sensor tag can provide reliable performance in wireless concrete-temperature measurement.

Acknowledgements

This work was supported by the National Natural Science Foundation of China (51664014, 51662008), the Key Research and Development Plan of Jiangxi Province (20161BBE50076) and the Science and Technology Project of the Education Department of Jiangxi Province (GJJ160491).

REFERENCES

Bicanic N and Zhang B (2006) Fracture energy of high-performance concrete at high temperatures up to 450°C: the effects of heating temperatures and testing conditions (hot and cold). *Magazine of Concrete Research* **58(5)**: 277–288, https://doi.org/10.1680/macr.2006.58.5.277.

Catarinucci L, De Donno D, Guadalupi M, Ricciato F and Tarricone L (2011) Performance analysis of passive UHF RFID tags with GNU-radio. In *Proceedings of the 2011 IEEE International Symposium on Antennas and Propagation (APSURSI), Spokane, WA, USA*, pp. 541–544.

Catarinucci L, Tedesco S, De Donno D and Tarricone L (2012) Platform-robust passive UHF RFID tags: a case-study in robotics. *Progress in Electromagnetics Research C* **30**: 27–39, http://dx.doi.org/10.2528/PIERC12042002.

De Donno D, Ricciato F and Tarricone L (2013a) Listening to tags: uplink RFID measurements with an open-source software-defined radio tool. *IEEE Transactions on Instrumentation & Measurement* **62(1)**: 109–118.

De Donno D, Catarinucci L and Tarricone L (2013b) Enabling self-powered autonomous wireless sensors with new-generation I^2C-RFID chips. In *Proceedings of the 2013 IEEE MTT-S International Microwave Symposium Digest (IMS), Seattle, WA, USA*, pp. 1–4.

De Donno D, Catarinucci L and Tarricone L (2014) A battery-assisted sensor-enhanced RFID tag enabling heterogeneous wireless sensor networks. *IEEE Sensors Journal* **14(4)**: 1048–1055.

Deng F, He YG, Li B *et al.* (2015a) Design of an embedded CMOS temperature sensor for passive RFID tag chips. *Sensors* **15(5)**: 11442–11453.

Deng F, He YG, Li B *et al.* (2015b) A CMOS pressure sensor tag chip for passive wireless applications. *Sensors* **15(3)**: 6872–6884.

Fu YF, Wong YL, Poon CS and Tang CA (2005) Stress–strain behavior of high-strength concrete at elevated temperatures. *Magazine of Concrete Research* **57(9)**: 535–544, https://doi.org/10.1680/macr.2005.57.9.535.

Kim S, Surek J and Baker-Jarvis J (2011) Electromagnetic metrology on concrete and corrosion. *Journal of Research of the National Institute of Standards & Technology* **116(3)**: 655–669.

Klysz G, Balayssac JP and Ferrières X (2008) Evaluation of dielectric properties of concrete by a numerical FDTD model of a GPR coupled antenna – parametric study. *NDT & E International* **41(8)**: 621–631.

Lei Z, Yi-Gang H, Isbella A, Yan-Qing Z and Ge-Feng F (2013) Analysis and measurements of path loss effects for ultra-high-frequency radio-frequency identification in real environments. *Acta Physica Sinica* **62(14)**: 144101.

LéGer P and Leclerc M (2007) Hydrostatic, temperature, time-displacement model for concrete dams. *Journal of Engineering Mechanics* **133(3)**: 267–277.

Lothenbach B, Winnefeld F, Wieland E, Alder C and Lunk P (2007) Effect of temperature on the pore solution, microstructure and hydration products of Portland cement pastes. *Cement & Concrete Research* **37(37)**: 483–491.

Schindler AK, Juenger MCG, Poole JL and Riding KA (2006) Evaluation of temperature prediction methods for mass concrete members. *Aci Materials Journal* **103(5)**: 357–365.

Tang WC and Lo TY (2009) Mechanical and fracture properties of normal- and high-strength concretes with fly ash after exposure to high temperatures. *Magazine of Concrete Research* **61(5)**: 323–330, https://doi.org/10.1680/macr.2008.00084.

Vyas R, Lakafosis V, Lee H *et al.* (2011) Inkjet printed, self-powered, wireless sensors for environmental, gas, and authentication-based sensing. *IEEE Sensors Journal* **11(12)**: 3139–3152.

Wang XT, Taylor P, Wang KJ and Lim M (2015) Monitoring of setting time of self-consolidating concrete using ultrasonic wave propagation method and other tools. *Magazine of Concrete Research* **68(3)**: 1–12, https://doi.org/10.1680/macr.15.00076.

Wei P, Che W, Bi Z *et al.* (2011) High-efficiency differential RF front-end for a Gen2 RFID tag. *IEEE Transactions on Circuits & Systems II Express Briefs* **58(4)**: 189–194.

Yan L, Xing Y and Li J (2013) High-temperature mechanical properties and microscopic analysis of hybrid-fibre-reinforced high-performance concrete. *Magazine of Concrete Research* **65(3)**: 139–147, https://doi.org/10.1680/macr.12.00034.

Zhang GX (2007) Temperature control of roller compacted concrete dam. *Water Resources & Hydropower Engineering* **17(7)**: 435–439.

Zhang R, Shi N and Huang D (2013) Influence of initial curing temperature on the long-term strength of concrete. *Magazine of Concrete Research* **65(6)**: 358–364, https://doi.org/10.1680/macr.12.00107.

Zuo L, He YG, Li B, Zhu YQ and Fang GF (2012) Analysis and improvement for UHF RFID reading region in real environments. *Acta Physica Sinica* **61(24)**: 244103.

Dhir and Paine
ISBN 978-0-7277-6457-7
https://doi.org/10.1680/icetsc.64577.237
ICE Publishing: All rights reserved

Chapter 15

Turning segmental tunnels into sources of renewable energy

Jan Niklas Franzius Dipl Ing, PhD, DIC, CEng, MICE
Team leader, Tunnel engineering department of Ed. Züblin AG in Germany

Norbert Pralle Dr-Ing, MA
Head of the research, Development and innovation department of Ed. Züblin AG
in Germany

Geothermal energy is a universal low-grade renewable energy source and can be built into any engineering structures that provide large interface areas with the ground. There are already a number of case studies on retrieving geothermal energy from tunnels lined with sprayed concrete, but this paper presents a new system for segmental tunnel linings. Following a field trial in Germany it has been successfully installed in a new high-speed railway tunnel in Austria to supply a municipal building with heating energy. The paper demonstrates how subsurface urban structures can contribute both to sustainable city planning and the provision of decentralised renewable energy.

Reducing carbon dioxide emissions has become an integral aspect of the design and operation of all constructed facilities. It can be achieved by more efficient use of resources and greater use of renewable or sustainable energy.

However, renewable energy sources such as wind, tidal, solar and geothermal are characterised by their decentralised nature whereas the greatest need for energy is in the world's ever-growing urban population centres. It is estimated that 35–40% of primary energy is consumed by built environments (Deutsche Bank Research, 2010) and that over 40% of carbon dioxide emissions relate to built environments (DTI, 2006).

Geothermal energy offers one of the best ways of providing decentralised sustainable energy in an urban environment, where the ground immediately below a city can be used as a low-grade energy storage reservoir. Subsurface structures such as piles, retaining walls and tunnels feature large structure–soil interfaces and can all be converted for use as heat-exchangers.

There have already been field trials and applications of geothermal systems in tunnels with sprayed concrete linings (Adam and Markiewicz, 2009; Schneider and Moormann, 2010). However, a large proportion of urban tunnels are constructed using tunnel boring machines (TBM) and segmental linings. This paper presents a new system which allows segmentally lined tunnels to be turned into sources of renewable heat energy for the buildings above them.

Shallow geothermal systems

Extracting low-grade geothermal energy from shallow depths is an increasingly popular and proven technology for the heating and cooling of buildings (Brandl, 2006; Deutsche Bank Research, 2010). Preene and Powrie (2009) provided an overview of conventional geothermal systems. Closed-loop systems exchange the temperature of the ground by circulating heat-exchange fluid through a loop of buried pipes.

Geothermal systems can be operated via a heat pump or used for direct heating or cooling. Heat pumps are typically used for the heating of buildings while direct heating is applicable for keeping traffic surfaces (e.g. pavements and railway platforms) free of snow and ice. Cooling systems often do not require a heat pump as they operate as direct-cooling systems. The seasonal change between heating and cooling is of particular interest since the ground may serve as an energy reservoir, storing superfluous heat from the summer to be used for heating during winter.

Heat pumps work in principle like a reversed refrigerator. A typical heat pump has a coefficient of performance – which is the ratio between energy output of a heat pump and the energy required to operate the system – in the range 3 to 4.

Tunnels for geothermal energy

There are a number of case studies on retrieving geothermal energy through tunnels. Early examples are reported from Switzerland (SVG, 2008) and in Austria, geothermal energy systems have already been installed in tunnels lined with sprayed concrete (Adam and Markiewicz, 2009).

Many new tunnels, however, are now being constructed using full-face TBMs. This technique enables tunnel excavation in weak ground with a minimal risk of damage to existing surface structures. It is therefore particularly advantageous for inner-city projects.

The lining of such tunnels usually consists of prefabricated concrete segments which form rings typically 1–2 m wide. Incorporating a heat-exchanger system into these segments has to meet the following requirements

- structural integrity of the lining must not be compromised
- watertightness of the lining must be maintained
- fitting of heat-exchange pipes should be integrated in the segment manufacturing process
- pipe connection between adjacent segments must not interfere with the TBM advance, but must be robust enough to fulfil durability requirements.

Heat-exchanger segments

The 'Energietübbing'® heat-exchanger segments developed by Ed. Züblin AG and Rehau AG & Co (Pralle *et al.*, 2009a) allows the conversion of segmentally lined tunnels into large heat-exchangers to harvest geothermal energy from the surrounding soil (Figure 1).

The segments are fabricated in the standard segmental lining manufacturing process. The only difference to standard segments is the integrated heat-exchange pipe (cross-linked polyethylene, approximately 20 mm diameter) which can be tied to the reinforcement cage (Figure 2) or a

Figure 1 Typical segmental tunnel showing a ring of Energietübbing heat-exchanger segments

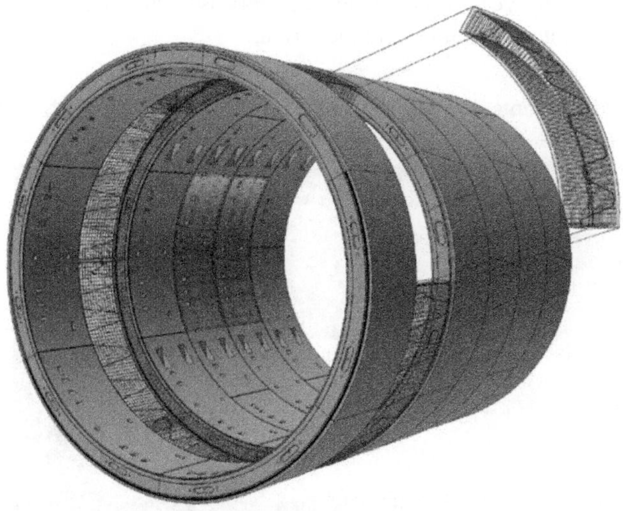

Figure 2 Concreting a segment fitted with a 20 mm diameter polyethylene heat-exchange pipe of 20–30 m length

simple frame structure in the case of fibre-reinforced concrete. The concrete cover is not compromised since the pipes are tied at the inside of the reinforcement cage. Pipe lengths of 20–30 m are placed into each segment in a meandering fashion.

The connecting ends of each segment's pipe face the tunnel's interior. Special moulds hold the ends during the casting and provide pockets which allow coupling of neighbouring heat-exchange pipes (Figure 3). The coupling requires high accuracy, but no more than that required for standard segment production.

The heat-exchange pipes are coupled after the segments have been installed to form a complete ring (Figure 4). The coupling system has been successfully used in standard heating and cooling systems for many years. The connecting work is performed from the TBM backup system so it does not interfere with the tunnel drive.

Figure 3 Completed segments showing coupling pockets at the end of each heat-exchange pipe

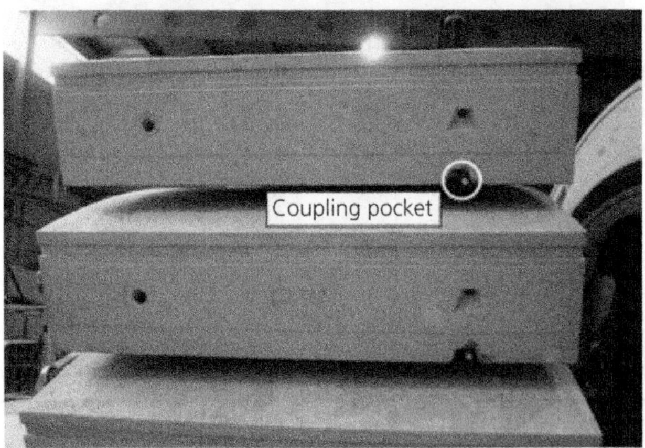

Figure 4 Completed tunnel lining showing connections between segment heat-exchange pipes

Modelling heat flux

A finite-difference calculation was performed to estimate the heat flux potential of tunnels equipped with heat-exchanger lining segments. The calculations were based on the following assumptions.

- The temperature field around tunnel is axi-symmetric and remains constant in the longitudinal direction of the tunnel.
- Primary heat transfer is through conduction; convection (the influence of groundwater) was not considered.
- The ground is isotropic and homogenous regarding thermal properties.

The corresponding differential equations are derived by many authors (e.g. Brandl, 2006). Based on these assumptions a simplified, axi-symmetric finite-difference model was adopted for the simulation (explicit forward time, centered space scheme; see for example Recktenwald, 2004). The thermal parameters used are summarised in Table 1.

The initial ground temperature was set to 12°C and held constant at a distance of 30 m to the tunnel lining. Three different cases are presented.

- Case 1: permanent operation of heat pump: heat is extracted from the ground continuously.
- Case 2: interval operation of heat pump: the heating pump only operates for a specified number of hours per day, changing with seasons. The geothermal system is not operated during July and August.
- Case 3: interval operation of heat pump combined with building cooling period during July and August.

A period of 3 years was simulated. The withdrawal of the heat was modelled by prescribing a heat flux of 10 W/m^2 (heat extraction) at the inside boundary of the tunnel lining. The heat flux was reversed when modelling the cooling phase during the summer. Figure 5 shows the resulting temperature evolution at the contact point between tunnel lining and soil and, for case 3, 5 m into the ground.

In case 1 the soil temperature drops drastically and reaches 0°C after approximately 3 months. In this case, the heat reservoir would be depleted within a short time period (as the effect of groundwater flow was disregarded).

Table 1 Input parameters for numerical analysis

Thermal conductivity, λ	3·3 W/m.K
Specific heat capacity, c	1100 J/kg.K
Density, ρ	2400 kg/m^3
Initial ground temperature	12°C

Figure 5 Modelled ground temperature for different operations of the tunnel heat-exchange system (all at lining–ground interface except where shown)

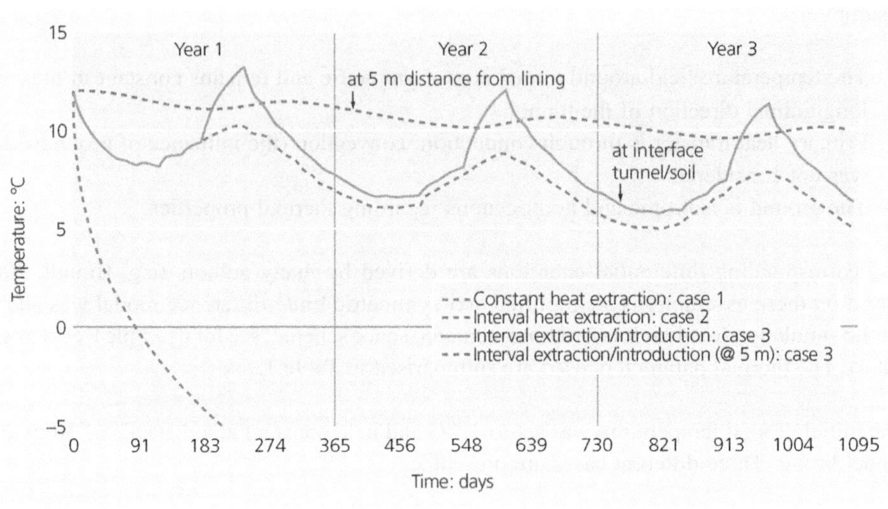

In case 2, the heat flux is only applied for several hours per day – from 7·2 h in January to 1·2 h in June and no heating during July and August. A total of approximately 1500 h of heating per year was modelled. The graph shows the temperature bouncing back during periods with few or no heating loads. This indicates a significantly slower heat extraction rate than in case 1 and keeps the temperature clearly above 0°C over the simulated time of 3 years (albeit there is a trend of reducing temperature).

In case 3, a negative heat flux of -10 W/m^2 was introduced into the ground during the summer months July and August. This represents cooling a building by charging heat into the ground. The recharging of heat during the summer months leads to higher temperatures in the following winter months and, consequently, the temperature in this case remains higher than in the second case with heating operation only. This scenario illustrates the benefit of using the ground as a seasonal store of heat energy.

Field trial in Germany

Four rings of a new high-speed railway tunnel in Germany were equipped with heat-exchanger segments in November 2007. They were subsequently activated for a field trial between May and September 2009 prior to the tunnel opening, using a temporary heat pump and monitoring system.

Five segments were temporarily connected to become a semi-circular activated surface, with an area of approximately 60 m^2 (Figure 6). The number and position of the segments used in the field trial was chosen according to the space requirements of the ongoing construction

Figure 6 Layout of field trial in the Katzenbergtunnel high-speed rail tunnel in Germany – five heat-exchanger segments were temporarily connected and operated during the construction period

works in the tunnel. The heat-exchange pipes were connected inside the tunnel (in contrast to the coupling system displayed in Figure 4) to fulfil the client's requirement to remove the installations after completion of the field trial.

The test was comprehensively monitored, featuring over 40 gauges to read temperatures within the segments (installed prior to casting), the tunnel climate, segment surfaces and up to 7·5 m into the surrounding rock.

The test consisted of the following phases.

- Phase 0: background temperature reading (approximately 3 months).
- Phase 1: thermal response test (4 days): heat charged into the ground through constant heat flux.
- Phase 2: no heat flux imposed: thermal relaxation of system.
- Phase 3: operation of heat pump (approximately 2·5 months): heat extracted from the ground.

Thermal response tests have become standard practice for the determination of the thermal conductivity of the ground during the design phase of conventional geothermal applications, such as energy piles. The phase 1 test was characterised by a constant power level (heating), which was imposed onto the system over a certain time period. The surrounding rock was charged throughout the test with a heat power level of 1·0 kW at first, which was then increased to 1·5 kW after 72 h. The test took 4 days.

A tunnel heat-exchanger is more complex than a conventional geothermal system, since heat is also transferred inside the tunnel. The results of the thermal response test therefore reflect the interaction between the hollow tunnel, the tunnel lining and the surrounding ground.

The phase 3 test consisted of cooling the heat-exchange fluid, which simulates a heating period during winter through extracting heat from the ground. A heat pump with power 1·0–1·5 kW was operated during this phase.

Disregarding the heat loss via the connecting pipes, the heat flux would equate to 17–25 W/m^2. When considering that the connecting pipes (outside the tunnel lining) contributed to the heat exchange, the heat flux from the surrounding ground is estimated to be in the range 10–20 W/m^2. Since the temperature in the heat-exchange fluid remained well above 0°C it is estimated that the heat flux could be further increased by operating the heat pump at a lower temperature level.

Figure 7 presents the temperature distribution along a temperature monitoring rod, reaching 7·5 m deep into the surrounding rock. The data indicate that the heat extraction affects the ground temperature up to a depth of 5 m.

One of the segments used in the field trial was also tested in a laboratory using thermal imaging (Figure 8). More details about the laboratory test are summarised by Pralle *et al.* (2009b).

Figure 7 Measured ground temperature during the field trial, which consisted of a 4 day heat-introduction phase, a relaxation phase and two 2·5 month heat-extraction phases

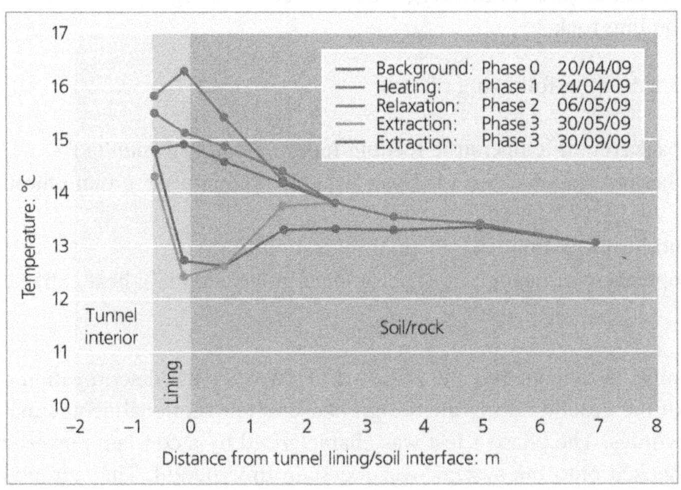

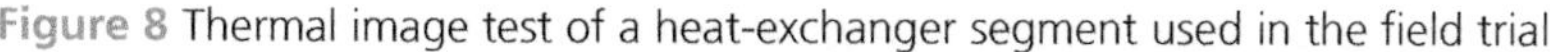

Figure 8 Thermal image test of a heat-exchanger segment used in the field trial

Demonstration project in Austria

Following the field trial, the geothermal system was installed as a demonstrator project in a 54 m length of a new twin-track high-speed railway tunnel in Austria to supply around 40 kW heat for a municipal building in the town of Jenbach above the tunnel. The 12 m diameter tunnel is in the Inn Valley and forms part of the northern approach route for the proposed Brenner Base tunnel. The local geology is dominated by gravel layers and overburden depth is approximately 27 m.

Each tunnel ring is formed by seven segments plus a key stone and is 2 m wide and 0·5 m thick. A total of 27 heat-exchanger segment rings were installed next to an escape adit and shaft through which the heat-exchange pipes are connected with the building heating system (Figure 9). The distance between the building and shaft is less than 150 m, and the heat pump and controls for the geothermal system are located within the building.

It should be noted that the heat-exchanger segments were designed after standard-segment production was already running. Reinforcement cages were only slightly modified and the size of coupling pockets required some optimisation. However, such issues could easily be eliminated if the heat-exchange system was considered in the segment design from the outset.

Figure 9 Layout of demonstrator project in Jenbach, Austria – water circulating in a 54 m length of a new 12 m diameter high-speed rail tunnel lining is pumped up an escape shaft to heat a nearby municipal building

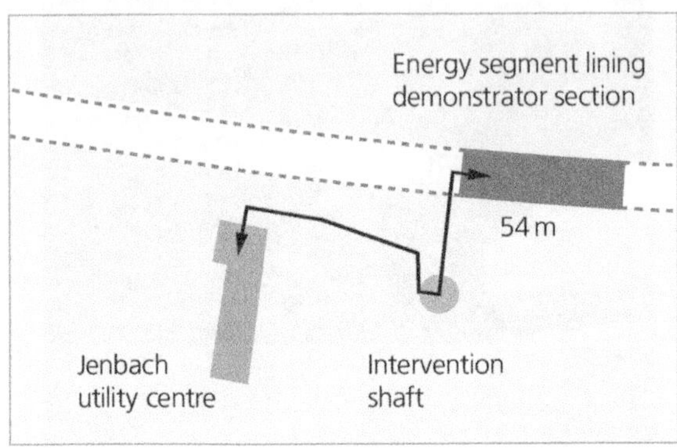

The heat-exchanger segments were successfully installed in November 2008. Coupling of adjacent segment pipes was carried out from the TBM back-up system and, after tunnelling was completed, the pipe circuits were connected along the invert to the escape shaft (Figure 10). During pressure testing, hot water was applied to illustrate the thermal effects on the lining of the system (Figure 11).

The building geothermal heating system is scheduled to become operational in April 2011.

Figure 10 Heat-exchange-pipe circuits were connected along the invert prior to construction of the tunnel base slab

Figure 11 Heat-exchange-pipe circuits in each ring were pressure-tested using hot water (a), the thermal effects of which can be seen with thermal imaging (b)

Capacity, yield and savings

The tunnel length required to supply a building's thermal energy depends on the heat flux and the building's requirements. The heat flux is mainly dependent on the tunnel geometry, the thermal ground properties and hydrogeological conditions.

The results from the field trial indicate a heat flux typically in the range 10–20 W/m^2. Table 2 presents the required length of a single 12 m diameter tunnel lined with heat-exchanger segments for various energy demands, and a heat pump coefficient of performance of 4. This indicates the demonstration project should be capable of supplying a heating demand of around 40 kW.

The carbon dioxide emission savings of a geothermally operated heating system compared with a conventional heating system depend on the energy mix used to power the heat pump and the type of conventional heating system which is to be replaced.

For conventional oil-fired boilers, carbon dioxide emission savings in the order of 55% are indicated by Schubert and Kaschenz (2008) and HPC (2008). The savings compared with gas-fired boilers are lower: Schubert and Kaschenz (2008) and HPC (2008) state approximately

Table 2 Required length of heat-exchanger lining depending on heat flux and heat demand

Heat demand: kW	Required tunnel length for different heat fluxes: m		
	10 W/m^2	15 W/m^2	20 W/m^2
20	41	27	20
40	82	55	41
60	124	82	61

25–35% reduction. These numbers are based on typical energy mixes producing carbon dioxide emissions of 470 g/kWh (EU average; see HPC, 2008) to 620 g/kWh (German average; see Schubert and Kaschenz, 2008).

The savings could be improved by reducing the carbon dioxide emissions of the electricity required to operate the heat pump. The Jenbach project, for example, is located in Austria, which features a higher proportion of renewable energy (of over 20% according to European Commission, 2007) compared with other European countries, so average carbon dioxide emissions are approximately 220 g/kWh. If the geothermal system was additionally used for cooling purposes, further reductions in carbon dioxide emissions would be achieved.

Conclusion and outlook

The development and implementation of a heat-exchanger segment lining system has shown that TBM-driven tunnels can be used for economically viable geothermal applications. Tunnels are designed to last for more than 100 years, and installing geothermal systems within them adds to their value.

Tunnels are routinely built as part of city and infrastructure planning. Additional investments for integrating the geothermal system are relatively minor compared to the costs of the tunnel construction, and offer an additional value for investment.

Even if the extra costs at current prices may not offer a return on investment within the normal investment horizon of 7–10 years (though the authors believe it would), large infrastructure projects can be treated differently as the structures will serve for many decades. There is a limit as to how much future electricity can be provided by renewables such as wind power, so the long design life of tunnels needs to be factored into today's energy planning.

The work shows that heat exchangers in tunnel structures are able to provide a reliable amount of renewable energy in places where the energy potential would otherwise not be exploited. The heat-exchanger lining segment system does not replace conventional geothermal applications, but enhances their field of applicability. It could also be sued to extract superfluous heat from sewers.

In addition, construction of new urban tunnels is also often linked to regeneration programmes along the tunnel route. This offers the opportunity to combine the design of new low-emission buildings with the supply of sustainable energy from the new tunnel.

Acknowledgements

The authors would like to thank all colleagues involved in the development of the heat-exchanger lining segments, specially the site teams of the tunnel projects Katzenbergtunnel in Germany and Jenbach H8 in Austria and their colleagues at Rehau.

REFERENCES

Adam D and Markiewicz R (2009) Energy from earth-coupled structures, foundations, tunnels and sewers. *Géotechnique* **59(3)**: 229–236.

Brandl H (2006) Energy foundations and other thermo-active ground structures. *Géotechnique* **56(2)**: 81–122.

Deutsche Bank Research (2010) *Geothermal Energy – Construction Industry a Beneficiary of Climate Change and Energy Scarcity.* Deutsche Bank Research, Frankfurt, Germany.

DTI (Department of Trade and Industry) (2006) *Our Energy Challenge: Power from the People, Microgeneration Strategy.* Department of Trade and Industry, London.

European Commission (2007) *Austria – Energy Mix Fact Sheet.* European Commission, Brussels, Belgium. See http://ec.europa.eu/energy/energy_policy/doc/factsheets/mix/mix_at_en.pdf (accessed 20/10/2010).

HPC (Heat Pump Centre) (2008) *Heat Pumps Can Cut Global Carbon Dioxide Emissions by 8%.* Heat Pump Centre, SP Technical Research Institute of Sweden, Borås, Sweden. See http://www-v2.sp.se/hpc/publ/HPCOrder/ViewDocument.aspx?RapportId=451 (accessed 20/10/2010).

Pralle N, Franzius JN and Gottschalk D (2009a) City district – mobility and energy supply: synergy potential of geothermal activated tunnels. *VDI Bautechnik* **84**: 98–103. (in German).

Pralle N, Franzius JN, Acosta F and Gottschalk D (2009b) Using tunnelling concrete segments as geothermal energy collectors. In *Innovative Concrete Technology in Practice, 5th Central European Congress on Concrete Engineering, Baden, Austria, 2009.* Austrian Society for Concrete and Construction Technology, Vienna, pp. 137–141.

Preene M and Powrie W (2009) Ground energy systems: from analysis to geotechnical design. *Géotechnique* **59(3)**: 261–271.

Recktenwald GW (2004) *Finite-difference Approximations to the Heat Equation.* Portland State University, Portland, OR, USA. See http://web.cecs.pdx.edu/~gerry/class/ME448/codes/FDheat.pdf (accessed 21/10/2010).

Schneider M and Moormann C (2010) GeoTU6 – a geothermal research project for tunnels. *Tunnel* **2**: 14–21.

Schubert J and Kaschenz H (2008) *Elektrische Wärmepumpen – eine erneuerbare Energie?* Umweltbundesamt, Dessau-Roßlau, Germany, Fachgebiet I 2.4 (in German). See http://www.umweltdaten.de/publikationen/fpdf-l/3192.pdf (accessed 21/10/2010).

SVG (Schweizerische Vereinigung für Geothermie) (2008) *Tunnelgeothermie – eine nutzenswerte Energiequelle im Land der Tunnels, Technische Notiz.* Schweizerische Vereinigung für Geothermie, Frauenfeld, Switzerland (in German). See http://www.geothermie.ch/data/dokumente/TechnischeNotizen/DE/Notiz4.pdf (accessed 21/10/2010).

Dhir and Paine
ISBN 978-0-7277-6457-7
https://doi.org/10.1680/icetsc.64577.251
ICE Publishing: All rights reserved

Index

Page numbers followed by f and t indicate figures and tables, respectively.